Environment, Development and Social Change in the Himalayan Region

Environment, Development and Social Change in the Himalayan Region

Sakarama Somayaji
&
Vimal Khawas

AKANSHA PUBLISHING HOUSE
NEW DELHI • GUWAHATI • VISAKHAPATNAM

AKANSHA PUBLISHING HOUSE
4649-B/21, Ansari Road
Darya Ganj, New Delhi - 110 002
Ph.: 23263193 / 9811582579 / 09435116718
Fax: 011-23263193
Email: ektabooks@yahoo.com
www.akanshapublishinghouse.com

Environment, Development and Social Change in the Himalayan Region

First Published 2012
ISBN 978-81-8370-319-2

PRINTED IN INDIA

Published by Akansha Publishing House, New Delhi and Printed at Harmain Offset, Delhi.

Contents

	Preface	vii
1.	Introducing the Himalaya: Environment and Human Dynamics *—Sakarama Somayaji & Vimal Khawas*	1
2.	Environmental Degradation, Climate Change and Disasters in the Eastern Himalayas *—Walter Fernandes*	22
3.	Vulnerability Assessment of Rangeland Dependent Mountain Communities Under Climate Change in the Hindu Kush-Himalayas (HKH) *—Jasra A.W., S. Joshi & M. Ismail*	29
4.	Climate Change and Large Cardamom Farming in the Sikkim Himalaya *—Vimal Khawas*	47
5.	Environmental Constraints and Population Pressure in a Mountain Ecosystem of Western Himalaya: A Case Study of Kinnaur *—Uttam Lal*	70
6.	Medicinal Plant Biodiversity in the Darjeeling Himalaya *—Dhani Raj Chhetri*	91
7.	Understanding Social and spatial Exclusion in India *—Sohel Firdos*	116

8. Migration from North Eastern Region to the Rest of the Country: A Recent Patterns and Trends 148
—Reimeingam Marchang

9. HIV/AIDS in Darjeeling: Historical Context and the Recent Scenario 169
—Binu Sundas

10. Right to Free and Compulsory Education: A Forthcoming Challenge for Sikkim 191
—Abdul Hannan

11. The Dynamics of Women, Work and Social Status: A Case Study of the Bhutias of Sikkim 211
—Swati Akshay Sachdeva

12. Cloth, Women and Social Change: Situating Handloom Weaving in Manipur 237
—Otojit Kshetrimayum

13. Traditional Healing Practices in Darjeeling Himalayan Region 273
—Santosh Kumar Rai

14. Present Status, Scope and Future Prospects of Floriculture in Uttarakhand 284
—Deepti Singh

Our Contributors 293

Index 296

Preface

Mountains form one of the most important bio-geographical resource zones of the world. They are remote areas covering 52 per cent of Asia, 36 per cent of North America, 25 per cent of Europe, 22 per cent of South America, 17 per cent of Australia, and 3 per cent of Africa, making up 24 per cent of the earth's continental surfaces. They encompass some of the most awe-inspiring landscapes, a great diversity of species and habitat types and distinctive, tenacious and often disadvantaged human beings. They directly support the 22 per cent of the world's people who live within mountain regions. A further 40 per cent live adjacent or very close to mountain areas and benefit from mountain resources in more than one ways. Well over half the global population depends on mountain environments for a wide range of goods and services including those of water, food, hydro-electricity, timber, biodiversity maintenance and mineral resources, besides availing opportunities for recreation and spiritual renewal. Up to 80 per cent of the planet's fresh surface water comes from the mountains. Hence, mountains are very significant to humans in a variety of ways. Scholars have often argued that the future security of the planet's growing human population rests in great measure on mountain watersheds.

Mountains are, however, fragile resource zones and are highly susceptible to both natural forces and anthropogenic factors. Mountain people face an environment where everyday physical demands are great, natural hazards are significant and agricultural production is constrained. Only about 3 per cent of land ranked as highly suitable for rain-fed agriculture is within mountains, highlighting the restricted livelihood opportunities available to many mountain people. Further, difficult access with socio-economic and political marginalisation often compounds the problems. In recent times many anthropogenic activities have been aggravating the natural setup of mountains.

The Himalaya, one of the youngest mountain systems in the world, provides us with very relevant examples with regard to the importance of environmental factors on the overall human security in mountain geo-environment. The Himalaya-Hindukush region is an integral part of the global ecosystem. The Himalayan geography, to varying degree, affects the climate, biodiversity, water resources, soil condition, socio-cultural pattern, economy and overall life and prosperity of the countries encompassed and their immediate surroundings.

As a major resource-zone of South Asia, it provides numerous goods and services to the people of both uplands and lowlands besides acting as a strong defence barrier against aggression by enemies, thereby, performing the role of a primary source of regional, environmental and human defense. This mountain region is rich in biological and environmental resources and serves as a water tower for the region, and the world. Nine Himalayan river systems flow along these ranges and provide direct basis for livelihoods for over 150 million people. In total, they sustain the lives of over 1.3 billion people – a fifth of the world's population. In fact, the importance of Himalaya in the context of regional environmental security and the resultant human wellbeing is beyond ordinary human perception.

The Himalaya is, however, not as strong as it appears to be. Dynamics of change in the entire Himalayan ranges are inextricably intertwined with factors such as climate, geology, fauna, flora, water resources etc. The region experiences various types of geo-environmental/geomorphic processes and thereby is prone to a number of natural hazards. Often these hazards assume the form of disaster due to the region's inherent nature, climatic conditions and lack of adequate disaster preparedness mechanisms, consequently inflicting widespread catastrophe and human insecurities in the region. However, more recently, many of the natural processes have been accelerated by unscientific and unsustainable anthropogenic activities.

Earthquakes ranging in severity, floods/flash floods, and glacial lake outbursts are common among other major natural hazards. Monsoon is the time when water-induced disasters take place in some form or other in various places across the Himalaya. Cloudbursts are generally reported every monsoon period wreaking tremendous environmental and human devastation within the region and downstream.

Most of the rivers in the Himalayan terrain flow through narrow gorges abutting moderate to steep slopes with sharp bands and meet tributaries on steeper slopes. As the rivers flow downstream, the valley becomes comparatively wider and less steep. The occurrence of flash floods, particularly in narrow river valleys, is one of the most feared consequences of major cloudbursts, landslides or glacial lake outburst. Rolling of debris by cloudburst or landslide along the constricted course of the rivers lead to a short-term damming of the river flow, resulting in the creation of temporary lakes, which can last anywhere from a few days to a few decades. When the backwater pressure of the lake exceeds the retention capacity of the barrier, the accumulated water gushes downstream with powerful force inundating otherwise safe settlements. Such flooding has an immense impact on the economy of the region and safety of the local population.

Majority of the settlements across the Himalaya are located on the middle slopes where fluvial terraces exist. People prefer to exploit such areas lying adjacent to the rivers. They are considered to be the most fertile locations by the Himalayan standards. However, such areas are also the most vulnerable to floods caused by the breaching of landslide-induced dams. Evidence of past damming in different river basins of the Himalayan region indicates that the river was blocked many times, particularly at the confluence with tributaries.

The fragile geology and torrential rains also play a significant role in destabilising the Himalayan terrain. The entire Himalayan belt is, both tectonically and seismically, a very sensitive domain with strong tectonised rocks and fragile mountain slopes vulnerable to the onslaughts of rains. The cumulative effects of past earthquakes in such a zone aggravate these phenomena. Even a cursory assessment of the severity of different natural disasters occurring in the Himalayan region indicates that the dwellers of this region are living with the great risk. Yet a comprehensive security mechanism in order to cope with these risks in the field has yet to be designed and developed.

Environmental insecurities across the Himalaya have been responsible in guiding and directing civilisations in the area over the millennia. The relationship between human and environment in the Himalaya had/has been very close since antiquity. In other words, environment and socio-economy are inseparable in the Himalaya. There had been a symbiotic and intimate relationship between humans and environment over the ages. However, more recently, things are changing for bad in this fragile resource zone. Originally, humans adjusted themselves with the complex geo-environmental setup of the region and hence environmental determinism was very strong. In recent times, however, we have been seeking to adjust the Himalayan environment according to our needs thereby making way for environmental possibilism in this otherwise susceptible region. It has resulted in the imbalances in the environment and aggravated the natural processes. This is a very serious issue having far reaching implications on the security of the Himalayan environment and human beings living therein in the days to come.

The Himalayas, often, regarded as the cradle of South Asian civilisation are at present suffering from various human onslaughts - in addition to numerous natural forces that have been acting and reacting in the region since the earth's geologic past - in the form of faulty development policies/ ventures, unprecedented population growth and degradation/ depletion of its rich natural resource bases. What is more, the situation is deteriorating with every passing day and the future of the Himalayas looks clearly grim, given the rate of onslaught the region is forced to tolerate. Further, the Himalayas have been a battleground between environmentalists/conservationists and commercial/vested interests for the last half century. Such a situation has only politicised pertinent environmental issues therein and hence has done little good to the region. In the process, the indigenous and tribal populations of the Himalayas have been sidelined and are often regarded as direct agents of 'Himalayan Environmental Degradation'

The major objectives of this volume are to debate, discuss and draw upon the attention of the policy planners on some of the critical themes on the contemporary environmental, development and societal issues of the South Asian mountains systems and its geographical milieu. The issues confronting the great Himalayas including the North Eastern Hills of India are given special emphasis in the book. The chapters written by the combination of young and experienced scholars with distinct expertise on the issues pertaining to the hills and mountains of South Asia delve deeper into very many critical dynamics of mountain environment and development in the context of the changing paradigm.

The first article by the editors introduces the Himalayan region in the context of its environmental setup and human dynamics therein. It navigates the importance of Himalayan geography and highlights that the Himalayas, to varying degree, affect the climate, biodiversity, water resources, soil condition, socio-cultural pattern, economy and overall life and prosperity of the countries encompassed and their immediate surroundings. Walter Fernandes, throws some light on the burning issues of environmental degradation, climate change and disasters in the Eastern Himalayas, in his brief and thought-provoking article, along with a few alternatives and solutions.

Jasra AW et al. assesses Vulnerability of the Rangeland Dependent Mountain Communities under Climate Change in the Hindu Kush-Himalayan (HKH) region. It attempts to provide meaningful insights regarding the status of pastoral livelihoods and household perceptions towards vulnerability in the HKH region. *Vimal Khawas* in his chapter presents evidences of changes in the climate patterns in the Sikkim Himalaya and consequent impact on the large cardamom farming in the region. He highlights the adaptation and mitigation measures taken by the farmers with the help of the government interventions as well as through their own traditional methods. *Uttam Lal* in his essay tries to bring forth environmental constraints and population pressure in the mountain ecosystem of Western Himalayas through the case study of the Kinnaur district of Himachal Pradesh. Medicinal Plant Biodiversity in Darjeeling Hills has been very clearly described and discussed by *Dhani Raj Chhetri* in his article.

Sohel Firdos in his paper titled 'Understanding Social and spatial exclusion in India' focuses on the conceptual, theoretical and practical aspects of social and special exclusion in India. He draws upon the works of various writings of acclaimed scholars including Amartya Sen to build his case in the context of India. *Reimeingam Marchang* takes us to the very pertinent issue of Migration from North Eastern India Region to the Rest of the Country. He outlines the socio-economic factors of migration in the context of northeast and analyses the trend and patterns of migration from the northeast region of India. He concludes with very valid insights about the same.

Binu Sundas reports HIV/AIDS situation in Darjeeling Hills both in the historical and current contexts. The hills and mountains particularly located in the northeastern parts of India have historically suffered from the dilemma of HIV/AIDS. Several factors have been cited by the scholars in this connection. Darjeeling Hills have been one of the bad victims of this disease and the situation has become worse in recent times.

Abdul Hannan discusses the issues related to the right to free and compulsory education in Sikkim. The dynamics of Bhutia women in the Sikkim Himalaya particularly their work and social status have been clearly brought to record by *Swati Akshay Sachdeva*. She concludes that, Bhutia women in almost every index of autonomy decision-making authority, mobility, access to and control over resources, experience freedom as such. *Otojit Kshetrimayum* in his paper "Cloth, Women and Social Change: Situating Handloom Weaving in Manipur" assesses the culture of weaving in Manipur and examine the factors responsible for the development of weaving from a traditional craft to a commercial industry.

The Himalaya, a region of great natural wealth and biological diversity, is also known for the traditional culture and healing practices. *Santosh Kumar Rai* has very coherently documented traditional healing practices in Darjeeling Himalayan Region. He says, the greater dependence of people on traditional system of medical treatment has a direct association with their habitats, which are situated among difficult natural barriers and constraints. The lack of modern hospital facilities and support from the government and the so-called civilized society has also forced them to follow their indigenous practices. The final article of the volume is presented by *Deepti Singh* wherein she attempts to examine the present status, scope and prospects of Floriculture in Uttarakhand region.

We sincerely hope and believe that the articles and essays in the volume shall ignite the mind of the students, scholars and decision-makers to the various pertinent contemporary developmental and environmental issues featuring the Himalayan region.

We are tremendously grateful to the authors for their contribution to the volume. Finally, we also thank Ms. Apranta Somayaji, a young student from Delhi University, for her painstaking proof reading work.

Sakarama Somayaji
Development Sociologist
New Delhi

Vimal Khawas
School of Policy Planning and Studies
Sikkim University, Gangtok

1

Introducing the Himalaya: Environment and Human Dynamics

Sakarama Somayaji & Vimal Khawas

The Himalaya, one of the youngest mountain systems in the world, provides us with very relevant examples with regard to the importance of environmental factors on the overall human wellbeing in mountain geo-environment. The Himalaya-Hidukush region is an integral part of the global ecosystem. The Himalayan geography, to varying degree, affects the climate, biodiversity, water resources, soil condition, sociocultural pattern, economy and overall life and prosperity of the countries encompassed and their immediate surroundings.

As a major resource-zone of South Asia, it provides numerous goods and services to the people of both uplands and lowlands besides acting as a strong defence barrier against aggression by enemies, thereby, performing the role of primary source of regional environmental and human security. This mountain region is rich in biological and environmental resources and serves as a water tower for the region, and the world. Nine Himalayan river systems flow along these ranges and provide direct basis for livelihoods for over 150 million people. In total, it sustains the lives of over 1.3 billion people – a fifth of the world's population. In fact, the importance of Himalaya in the context of regional environmental security and the resultant human-wellbeing is beyond ordinary human perception.

Ecological Setup and Environmental Resource Base

The Himalaya has with it the most striking and distinctive features on the earth's surface and effects to varying degree the climate, water resource, soil condition, biodiversity, economy and overall life and prosperity of the countries encompassed.

Climate and Natural Vegetation

Himalaya exhibits one of the most complex climatic patterns and it is virtually far from possible to come to a clear pattern of climatic set up in the region. Owing to its extreme diversity in its various sub-ecological locations, elevation, and topography the region is featured by extreme variants of climate. Two main climatic characteristics of the region are the seasonal rhythm of weather and the vertical zoning. The climatic conditions vary from hot sub-humid tropical in the southern low tracts to temperate, cold alpine and glacial in the northern high mountains.

The temperature of the region varies spatially—both vertically and horizontally. April-May provides the highest temperatures, with maximum exceeding 40°C at many lowland stations (Ives and Messerli 1989) while mean monthly temperature is lowest in January (Singh 1971) deeping over –20°C in much of the higher Himalaya. The lowest recorded temperature in this connection has been recorded in Leh with –28.3° C while exceptionally high temperature has been recorded in Jammu with 47.2° C (Singh 1971). As a whole, climate ranges from subtropical in the southern foothills, with average summer temperatures of about 30° C and average winter temperatures of about 18° C; warm temperate conditions in the Middle Himalayan valleys, with average summer temperatures of about 25° C and cooler winters; cool temperate conditions in the higher parts of the Middle Himalayas, where average summer temperatures are 15 to 18° C and winters are below freezing; to a cold alpine climate at higher elevations, where summers are cool and winters severe (Karan [undated]). At elevations above 4880 m the climate is very cold with below freezing temperatures and the area is permanently covered with snow and ice (Karan [undated]).

Of the total annual precipitation, 70-85 per cent falls during June-September, depending upon location (Ives and Messerli 1989). According to available records the highest precipitation occur between the Annapurna massif (Ibid) and the eastern Himalaya. The highest amounts are recorded north of the Bay of Bengal in Darjeeling-Sikkim, Assam Hills and Arunachal Pradesh.[1] With increasing distance along the Himalayan front toward the west-northwest, total annual precipitation decreases and the occasional winter westerly disturbances become more important (Ives and Messerli 1989). The lower valleys and gorges are very dry and local agriculture is dependent upon snowmelt and glacial-melt irrigation commonly called *Kuhl* in Himachal Pradesh. Summer monsoon influences here are slight or absent. This over-all trend was noted by Troll (1938, 1939) while mapping altitudinal vegetation transects in the Nanga Parbat area (Ives and Messerli 1989). Troll's work was greatly expanded by Schweinfurth (1957), who produced the first composite vegetation map for the entire Himalaya. This map, a fundamental research resource to this day, reflects the parallelism between the climate and vegetation trends from north to south and from east to west (*Ibid*).

---

[1] Excluding the Northeastern hills.

This same pattern is also important in terms of the region's glacio-hydrology and the associated variation in water source and availability throughout the year (Young 1982, Ives and Messerli 1989). The eastern Himalaya, in general, provides moisture surpluses from direct runoff of the abundant summer monsoon rainfall; the snowmelt contribution is comparatively insignificant. With increasing distance toward the west-northwest, melt water becomes critically important. A particularly heavy summer monsoon, for instance, which produces excess water (and flooding) in the eastern half of the region, may only serve to lower the summer flow of the western rivers since the increased cloud cover (with little or no rain/snow) will serve to reduce incoming solar radiation and thus limit melt water production (Ives and Messerli 1989). A general lack of systematic studies in glacio-hydrology is observed to be a serious deficiency in view of the great importance attached to hydroelectric and irrigation potentials of the region.

The importance of the Himalaya in directing overall climatic pattern in and around the sub-continent may briefly be summerised as follows: It has a profound effect on the climate of the Indian subcontinent and the Tibetan Plateau. It prevents frigid, dry Arctic winds from blowing south into the subcontinent, which keeps South Asia much warmer than corresponding temperate regions in the other continents. It also forms a barrier for the monsoon winds, keeping them from traveling northwards, and causing heavy rainfall in the Terai region. The Himalaya is also believed to play an important part in the formation of Central Asian deserts such as the Taklamakan and Gobi deserts. Further, due to the presence of mighty mountain ranges, the western disturbances that appear from Iran during winter are prevented from traveling any further, resulting in snow in Kashmir and rainfall for parts of Punjab and northern India. Despite being a barrier to the cold northerly winter winds, the Brahmaputra valley receives part of the frigid winds, thus lowering the temperature in the Northeastern Indian states and Bangladesh. These winds also cause the north-east monsoon during this season in these parts.

Biogeography of the Himalaya

Biotic Province	*Biomes*	*Forest Types*
Western Himalaya	Tundra Zone Alpine Zone Temperate Zone Sub-Tropical Zone	Cold Temperate and Alpine Forest (2400-3600m) Warm Temperate Forests (1500-2400)
Central Himalaya	Alpine Zone Temperate Zone Sub-Tropical Zone	Sub-Tropical Forests (750-1500 m)
Eastern Himalaya	Alpine Zone Temperate Zone Sub-Tropical Zone	Tropical Forests (up to 900 m)

Source: Adopted (with some modifications) from Bir 1993, Khoshoo 1993.

Distribution of natural vegetation in the Himalaya has to be seen in the context of its biogeography. Rodgers (1985) classified the Himalaya as a boreal zone and identified several biomes within it. Accordingly, five major types of forest can be identified across the biomes of the Himalaya. They are Alpine forest, cold temperate forest, warm temperate forest, sub-tropical forest, and tropical forest. With the increase in elevation and microclimate, change in the vegetation type is observed. Further, as one moves from the west towards eastern flanks of the Himalaya, differentiation in nature and pattern of the vegetation is observed.

Natural vegetation belts range from tropical monsoon rain forest (*Shorea robusta* = sal forests) in the south, through a series of forest belts, to the upper timberline at approximately 4,000 4,500 m (Ives and Messerli 1989). Above this a rhododendron-shrub belt gives out onto alpine meadows, a sub-rival belt of extensive bare ground and scattered dwarf plants, mosses, and lichens, and finally, at 5,000-5,500 m, permanent ice and snow with steep rock outcrops (Ibid). Ives and Messerli (1989) following Joshi (1986) and Numata (1981) give the following generalised pattern of vegetation belts for the central part of Nepal.

The western Himalayan ranges are much wider and colder with drier climate, while the eastern ranges are among the wettest regions in the world (Khoshoo, 1993) with tremendous biodiversity. Consequently, the western ranges have vegetation that is cold loving and drought resistant. Large gregarious populations of conifers like chir and blue pines, deodar, fir and spruce feature the region. As we move toward the east conifers are not a dominant species, although they have their presence. Rhododendrons dominate the plant life of the region (Ibid). The difference in the distribution of vegetation across spaces of the Himalaya is the result of a number of factors. For instance, the western ranges lie at 36-degree north latitude and are an area of low rainfall affecting the tree line that is at 3600 m altitude. At the other extreme the eastern ranges lie at 27-degree north latitude and are among the wettest regions in the world. Tree line is higher at 4570 m (*ibid*). As a result, the western Himalaya is species deficient while eastern Himalaya is species rich. The average figure of total forest cover in the Himalaya (including Nepal, Bhutan and India) in relation to its geographical area comes to around 35 per cent.[2]

Water Resource

The Himalaya is often regarded as 'water tower' of South Asia. It is the storehouse of fresh water and source of mighty rivers like Indus, Ganges and Brahmaputra. These antecedent rivers that originate from the great glaciers of the Himalaya and Trans-Himalaya travel south of the Himalaya cutting across deep gorges, valleys, and hills

[2] The figure is adopted after examining the forest related data of the relevant countries.

defying the Himalayan orogeny. Numerous smaller rivers and streams join them, majority of which are glacier fed, perennial and antecedent in nature. They form three major rivers systems of South Asia namely, Indus river system, Ganges river system and Brahmaputra river system.

The Ganges system includes the catchment of the enormous Ganga river system that embraces much of the Himalaya, as far as West Bengal in the east, a densely populated Gangetic plain of India, and a part of Bangladesh. The western-most section of the Himalaya is drained by the Jhelum, Chenab, and Sutlej, major tributaries of the Indus, as well as by the upper Indus itself. The Brahmaputra is the Hindu name for that section of the other main trans-Himalayan river (Yarlungtsangpo-Brahmaputra) from the point where it enters Indian Territory. It rises high on the Tibetan Plateau in longitude 82° east, remarkably close to the main headstream of the Indus, and flows eastward for more than 1,200 km north of the Himalayan crest-line before making its spectacular turn to cut through the mountains in one of the world's most impressive gorges (Ives and Messerli 1989). It enters India through Arunachal Pradesh and flows roughly westward across Assam and Bangladesh before turning south again to produce a network of distributaries that merge with those of the Ganges, and eventually enters the Bay of Bengal through the great delta.

The Himalaya has approximately 23000-km^2 areas under glaciers making it one of the largest concentrations of glacier-stored water outside the Polar Regions (Sharma 2004:322). It stores about 5000 km^3 of permanent snow and ice (Upadhyay 1995). While glaciers cover 10-20 per cent of the total surface area of high mountains in Himalaya, 30-40 per cent has seasonal snow cover (ibid). Hence, the melt water has a great contribution to overall water yields in the river systems. However, the importance of melt water contribution diminishes from cold and arid western Himalaya to warm and humid eastern counterpart. Accordingly, the contribution of snow to the runoff of major rivers of the Eastern Himalaya is about 10 per cent (Sharma 1993, 2004) and more than 60 per cent in the Western Himalaya (Vohra 1981, Sharma 2004). The more availability of water in mountains due to orographic lifting of moisture is evident in terms of higher specific water yields for mountain watershed compared to whole river basins. The melt water forms important source of water in the north Indian rivers during critical summer months. The distribution of runoff produced from ice, snow and rain is such in the Himalayan Rivers that flow is observed in them round the year. The water yield from the high Himalayan catchments are roughly double the yields from the same size of catchments in peninsular India due to additional snow and ice melt contributions (Sharma 2004).

Water resource of major river system in the Himalayan Region

Category	*Indus*	*Ganges*	*Brahmaputra*
Water resource potential (km^3)	73.3	525	537.2
Utilisable surface water (km^3)	46.0	250	24.0
Groundwater potential (km^3)	25.5	171.7	27.9
Per capita annual availability of water (m^3)	1757	1473	18417
Per hectare of culturable area annual availability (m^3)	7600	8727	44232

Source: CWC 1993

Specific water yield from mountainous and whole river basin (mm)

River Basin	*Mountainous watershed*	*Whole basin*
Indus	460	163
Ganges	975	473
Brahmaputra	1039	922

Source: Alford 1985.

Land/Soil Resource

The diversity in its geological, physiographic, climatic and vegetation characteristics have given rise to a variety of soil types in the Himalaya. The formation of soils is the result of climatic factors and geomorphic processes supported by geo-lithology. The soil cover becomes thinner as we move higher in elevation towards alpine climate. Hence, thick soil covers are confined only to valley bottoms and terraces. As per the soil maps on 1:250,000 developed under a national project on Soil Resource Mapping of different states of India, majority of the Himalayan soils belong to Entisols and Inceptisols. The former are dominant in the Western and Central Himalayan belt while both are more balanced in their distribution in the Eastern Himalaya. In Northeastern Hills (Purvachal) the latter are found extensively. The Ultisols too occupy a considerable area of Purvachal.

Soils of the Himalayan Region (Area '000 Ha)

States	*Alfisols*	*Entisols*	*Inceptisols*	*Ultisols*	*Miscell-aneous*
Jammu & Kashmir	100.0 (0.5)	7572.6 (34.1)	147.2 (6.4)	-	47.7 (0.2)
Himachal Pradesh	238.0 (0.4)	2584.3 (51.3)	1099.9 (19.8)	-	400.0 (0.8)
Uttaranchal	8.8 (0.2)	3026.3 (56.5)	1623.2 (30.3)	-	11.1 (0.21)
Sikkim	-	231.7 (42.5)	239.8 (42.9)	-	816.5 (14.6)
Arunachal Pr.	25.1 (0.3)	2981.2 (35.6)	3123.6 (37.3)	1189.1 (14.2)	1055.1 (12.6)

Source: NBSSLUP 2004

Cultural Setting

Complicated, as the physical geography of Himalayan region is, the present-day cultural setting, economic and ethnic patterns resist easy description. This in turn is partly due to the influence of the physical base and partly a result of the very long and complicated history (Ives and Messerli 1989). However, human settlement and economic pattern in the Himalayan region are greatly influenced by the physiography/topography and climatic condition. The climatic provision is, in fact, one of the main factors that have directed population settlements across the region since historic past. Extreme climatic condition has in many cases imposed restrictions on the living conditions and movement/communication of the people.

Knowledge with regard to the pattern and extent of prehistoric settlement in the Himalaya, based upon archaeological evidence, remains very sparse. Consequently, attempts to reconstruct the social and intellectual forces that have shaped contemporary society in northern India and the Himalaya rely heavily upon the great body of Sanskrit epic literature, the earliest fragments of which date from 1200 BC (Ives and Messerli 1989). Much of this literature is derived from an oral tradition that had been continually modified over several thousand years before being committed to writing (O'Flaherty 1975, Ives and Messerli 1989). The history of India, especially the Hindu epics and Puranas tell us that the original inhabitants of the Himalaya were the Kinnars, Kilinds, and Kiratas. History[3] also mentions the names of Khasas and the Darads inhabiting the Himalayan region. The evolution of present human settlements across the Himalaya is, however, not easy to reconstruct as the information in this regard is scarce and often fragmented. However as of today, three different macro social groups, mainly, form the Himalayan population. They are the Negroids, Mongoloids and the Aryans. Ives and Messerli (1989) thus summarises:

[3] Readers may refer to Ives and Messerli 1989 for detailed history of the region.

> 'This somewhat fragmentary 'history' of settlement can be summed up with the statement that the northwestern part of the region evolved under Muslim influence, the southern flanks of the centre and east under Hindu influence, and the northern fringe under Buddhist influence. This broadly sweeping overlay conceals innumerable small ethnic groupings and says little about the independent entities such as Hunza, the great complexity of settlement of the Arunachal Pradesh Himalaya, and the several dozen distinct ethnic and linguistic groups of the Hengduan Mountains, which were eventually infiltrated along the main valleys by Han agricultural settlers and traders'.

Looking from south to north, the Hindus of Indian origin mainly dominate the Sub-Himalayan and Middle Himalayan valleys. While in the Great Himalayan region, it is mainly the Tibetan Buddhists who dominate the scene. From west to east the geographical distribution of the population may be summarised as follows– in places like eastern Kashmir and Nepal it is mostly Hindu population that dominate the area. However, in central Nepal, both Indian (mainly Hindu) and Tibetan cultures have blend together, producing a mixed culture of Indian and Tibetan traits. In the eastern Himalaya human diversity is much pronounced. Darjeeling-Sikkim region is a mixture of both Hindu and Tibetan culture while Bhutan has historically been a region of Tibetan culture. The Arunachal region of eastern Himalaya, along with some of its neighbouring hills, reflects the religion and culture similar to those living in Yunnan province in China while some of the northeastern states of India practice religion and culture of northern Myanmar. Further, a notable proportion of (converted) Christians also live in Darjeeling Hills, Arunachal Pradesh and neighbouring North-East Indian Hills. Muslims are mostly seen in western Kashmir and their culture is similar to the population of Pakistan, Iran and Afghanistan.

Human population of the Himalayan region, as of today, crosses 65 million (Khawas, 2009). They are mainly scattered across the lower and middle Himalayan region. The present ethnic pattern of the human settlement in the Himalaya has to be seen in the context of different political entities therein. Table [give the number….] summarises general ethnic composition of the Himalaya.

Major ethnic compositions in the Himalaya

Region	*Major Ethnic Community*
Bhutan	Bhote, Ethnic Nepali, Indigenous and migrant groups
Nepal	Chettri, Brahmin, Magar, Tamang, Newar, Kami, Yadav, Muslim, others
Jammu and Kashmir	Dogra, Gujjar, Gaddi, Kashmiri Pundit (Brahmin), Sunni- Shia- Hanji and Dard Muslims, Baltis, Ladakhi, others
Himachal Pradesh	Rajput, Brahmin, Ghirat, Mahajan, Sood, Chahang, Saini, Air, Darni, Lohar, Tarkhan, Nai, Dusali, Doomna, Chamar, Julaha, others
Uttaranchal	Kol or Kolta, Rajputs, Brahmin, Jaunsari, Bhotia, Buksha, Tharu, others
Darjeeling	Lepcha, Ethnic Nepali, Ethnic Bhutia/ Tibetan, others
Sikkim	Lepcha, Ethnic Nepali, Ethnic Bhutia/ Tibetan, others
Arunachal Pradesh	Monpa, Sherdukpen, Memba, Khamba, Khampti, Singpho, Adi, Aka, Apatani, Bangni, Nishing, Mishmi, Miji, Tangsa, Nocte, Wancho, others

Source: Compiled by the author.

Majority of the indigenous groups living in remote valleys and slopes of the Himalayan region have generally conserved their traditional cultural identities and institutions. However, often scattered and gradual improvements in communication and transportation system are slowly improving the lifestyle of the people living in many locations, mainly urban areas, of the Himalaya. Penetration of modernisation and associated forces has led to the gradual but steady deterioration of traditional-cultural and social systems in such locations.

Economic Arrangement

Agriculture and animal husbandry is the mainstay of the Himalayan population engaging large chunk of them in this sector. Agriculture is the primary driver of the economy.[4] Over 85 per cent of the population is directly or indirectly dependent on agriculture for its livelihood. The varied topographic and agro-climatic conditions permit the cultivation of a wide variety of crops and fruits ranging from sub-tropical to cool temperate. The cropping pattern reflects the varying conditions of altitude, climate and soil on the one hand and on other the various agronomic and cultural practices and traditions evolved over the centuries such as elaborative terracing, crop rotation and very efficient traditional irrigation system.[5] The region is also ideal for horticulture and fruit orchards.

The Himalayan region has been the source of several species of cereals, pulses, fruits, oil yielding plants, spices and tuberous vegetables and sugar yielding plants and their wild relatives. Added to it is a whole range of medicinal and aromatic plants some of which have gone into commerce and have their centre of origin in the Himalaya. Bamboo wealth of the Himalaya, particularly eastern region, needs special mention. Many species of bamboo extend up to western Himalaya, except Kashmir Valley. The Himalayan region harbours many wild and domesticated animal species whose abode falls on this region.

---

[4] Jhum, the local word for shifting cultivation in Eastern Himalaya, which was widely practised in Arunachal Pradesh among the tribal groups, has come to be less practiced.

[5] Such irrigation system is called *Kuhl* in Himachal, *Gul* in Uttaranchal and *Kulo* in Darjeeling-Sikkim. Further, in regions like Uttaranchal and Himachal Pradesh it also provides water for small water mills.

Agriculture is concentrated mainly on the Siwaliks and on the valleys of the Middle Himalaya. At some cases, however, patches of agricultural land have also been carved out in the high mountain areas. Rice is the principal crop in the well-watered valleys while maize is also an important rain-fed crop on the hillsides. Other cereal crops are wheat, millet, barley, and buckwheat. Sugarcane, tea, oilseeds, and potatoes are other major crops. Mixed farming is widely practiced in the Himalayan region. The general crop production association under mixed farming across diverse locations and elevations in the Himalaya are: 'barley-wheat-pulses, wheat-maize-rice, wheat-rice-ragi, maize-wheat-rice, maize-millet-rice, rice-millet-rice'.

Economic arrangement in the Himalaya

Region	*Major Economic Activity*	*Other Sources of Economy*
Jammu & Kashmir	Agriculture and allied activity (about 70%)	Household industry, Manufacturing, Tourism
Himachal Pradesh	Agriculture and allied activity (about 90%)	Forest, Hydropower, Tourism
Uttaranchal	Agriculture and allied activity (more than 80%)	Forest, Hydropower, Tea, Tourism, Small scale & cottage industry
Darjeeling	Agriculture and allied activity (about 26%)	Tea, Tourism, Forest
Sikkim	Agriculture and allied activity (more than 85%)	Hydropower, Tea, Tourism, Forest,
Arunachal Pradesh	Agriculture and allied activity (more than 90%)	Forest, Hydropower, Tourism, Cottage industry & handicrafts
Nepal	Agriculture and allied activity (more than 91%)	Forest, Hydropower, Tourism, Manufacturing, Household Industry.
Bhutan	Agriculture & allied activity (93%)	Forest, Hydropower, Tourism

Source: Compiled by the author.

However, Food production in the Himalaya has not kept up with the population growth. Agriculture is handicapped by steep and hilly terrain, climatic hazards, small and scattered holdings, thin stony soils, limited irrigation and small cultivated area. The scope for increasing cultivated area is not much. Physical environment restricts the cultivated land to a large extent in the area hence exhibiting acute shortage of arable land. It has further infused a high degree of adaptive skill and uncommon physical vigour among the people to cope with the inhospitality of the environment. To secure agricultural land, rugged terrains are arduously terraced even on such steep slopes where the land should be left under pasture or forest. 'Since alternative sources of livelihood other than agriculture are limited, the peasantry has to eke out its existence under difficult conditions where nature is a stern mother' (Singh 1971: 412).

Arable land and major agricultural activity

Region	*Arable Land (%)*	*Major Agricultural activity*
Jammu & Kashmir	5.0	rice, maize, wheat, millet, pulses, fruits, vegetable, saffron, animal husbandry
Himachal Pradesh	10.0	wheat, maize, rice, barley, pulses, millet, gram, potato, fruits, ginger, vegetables, *kuth*, sugarcane, medicinal herbs, animal husbandry, poultry, fish culture
Uttaranchal	14.4	millet, wheat, rice, ginger, sugarcane, fruits, vegetables, animal husbandry
Darjeeling	25.0	rice, maize, wheat, barley, millet, potato, fruit, vegetable, flowers, medicinal herbs, ginger, cardamom, livestock.

(Contd...)

Region	*Arable Land (%)*	*Major Agricultural activity*
Sikkim	11.0	rice, maize, cardamom, wheat, barley, millet, potato, fruits, vegetables, medicinal herbs, ginger, livestock.
Arunachal Pradesh	2.5	barley, wheat, maize, soybean, paddy, potato, millet, oilseeds, fruits, vegetables, spices, sugarcane
Nepal	20.2	rice, maize, wheat, barley, millet, oilseed, potato, jute, tobacco, cotton, indigo, opium, vegetable, fruits, medicinal herbs, livestock.
Bhutan	3.4	rice, maize, buck-wheat, root crops, citrus fruits, cardamom, spices, animal husbandry, poultry.

Source: Compiled by the author.

Besides, animal husbandry engages a sizeable proportion of the population in the Himalaya. This is particularly because of the fact that in a predominantly agrarian economy animal husbandry plays a vital role for supporting agricultural operations, for supplement food and for providing means of transport and sources of manure. Across the societies of the Himalaya large number of cattle, sheep, and goats are kept. However, they do not receive adequate nutrition (Singh 1971).

Irrigated area in the Himalaya

Region	*Net cultivated area (sq. km)*	*Irrigated area (%)*
Uttaranchal	662.4	33.9
Himachal Pradesh	580.0	17.9
Nepal	-	38.2
Bhutan	-	25.0

Source: Compiled by the author.

Himalaya, although has with it sufficient resource potential, is one of the least industrialised regions in the world. Very small per centage (less than 10 %) of Himalayan population is engaged in industrial and its associated sector. Remoteness, unfavourable geographical conditions, inadequate transportation and related infrastructure facilities, and unique historical forces have played prominent role in the regard.

The region has a large hydel power potential but we have not been able to tap this resource adequately. Further, although in recent times several hydel projects are under their ways there have often been stiff resistance from environmental groups and academia in view of the unsustainable ways water resource is tapped.

Apart from that, small-scale and rural industries have developed in several parts of the Himalaya and have the potential to grow in future. Rich and diversified histories of the region and the resultant socio-cultural mosaic have led the way to diversify traditional rural cottage industries. Several kinds of handicrafts and such other cultural items contribute significantly to the economy of the region.[6]

Tourism is yet another area where Himalaya has a comparative advantage. It is a growing industrial sector and it is growing relatively faster. However, it is surprising to know that the region remained isolated for long. Tourism is not a very old phenomenon in the area. It picked up only after the 1950s but it picked up relatively fast. This sector is yet to be properly regulated and efficiently diversified. Of late massive mass tourism pouring across spaces of the Himalaya coupled with weak arrangements for regulatory mechanism and inadequate institutions have been the serious cause of concern. Diversification of the sector into nature

[6] A visit to Delhi *Hat* of New Delhi provides one a good example in this connection.

tourism, eco-tourism, adventure tourism, cultural tourism, religious tourism, etc. is a welcome step if rationally planned and scientifically managed.

As a whole, the economy of the Himalaya is poor with low per capita income. Most of the population is dependent on agriculture, primarily subsistence agriculture as modern industries are lacking. Agricultural land is mainly concentrated in the foothills and valleys. Food production in the Himalaya has not kept up with the population growth (Karan [undated]). Recently, however, trade and commerce had played a vital role in the lives of the people living in frontier villages of the region.[7] Although strategic importance of the northern frontiers has accelerated the development of transportation and communication in the region after 1962[8] the isolation in most of the parts is yet to be broken. Commercial crops including horticulture and floriculture have been stimulated wherever modern roads and such other infrastructure facilities have reached.

Conclusion

Himalayan geography, to varying degree, affects the climate, biodiversity, water resources, soil condition, socio-cultural pattern, economy and overall life and prosperity of the countries encompassed and their immediate surroundings. As a major resource-zone of South Asia it provides numerous goods and services to the people of both the uplands and the lowlands and thus performs the role of prime source of regional environment and human security. The importance of Himalaya in the context of regional environmental security and the resultant human-wellbeing is beyond ordinary human perception. The region is crucial for the people living both inside the region and vast lowlands spreading across South Asia.

---

[7] Mainly in Himachal Pradesh, Ladakh, Kumaon and Garhwal.

[8] The Indo-Chinese War of 1962 accelerated the strategic importance of the Himalaya.

References

Bahadur, J. (1972), "The Himalayan Glaciers", *Science Today*, July: 21-27, 1972

Bahadur, Jagdish (2004), *Himalayan Snow and Glaciers: Associated Environmental Problems, Progress and Prospects*, New Delhi: Concept Publishing Company.

Bir, S.S. (1993), "Uniqueness of the Pterridophytic flora of the Himalayas and Conservation of Threatened elements", in U. Dhar (ed.) *Himalayan Biodiversity: Conservation Strategies*, Nainital: G.B.Pant Institute of Himalayan Environment and Development.

CEE Himalaya (2002), Sustainable Development in the Himalaya: Environmental Education and Communication Initiative, [Online: web] Accessed 8 September 2005, URL: www.mtnforum.org.

Central Bureau of Statistics (1998), *A Compendium on Environment Statistics 1998*, Central Bureau of Statistics: Kathmandu.

Central Bureau of Statistics (2004) *Handbook of Environment Statistics 2003,* Central Bureau of Statistics: Kathmandu.

CWC (1993), *Reassessment of Water Resource*, Central Water Commission, Govt. of India: New Delhi

Das, S T. (1978), *People of the Eastern Himalayas*, New Delhi: Sagar

Dhar, U. (ed.) (1993), *Himalayan Biodiversity: Conservation Strategies,* Nainital: G.B. Pant Institute of Himalayan Environment and Development.

Ives, J.D. (1986), Glacial Lake Outburst Floods and Risk Engineering in the Himalaya, *Occassional Paper No. 5*, ICIMOD: Kathmandu

Ives, J.D. (2002), "Along a steep pathway", Our Planet, [Online: Web] Accessed 15 May 2006, URL: http://www.ourplanet.com [UNEP]

Ives, J.D. (2004), *Himalayan perception: Environmental change and the well-being of mountain peoples,* London: Routledge, reprint 2006.

Ives, J.D. (2005), "Himalayan misconception and distortions: What are the facts", *Himalayan Journal of Sciences*, 3 (5): 15-24

Ives, J. D. and B. Messerli (1989), *The Himalayan Dilemma: Reconciling development and conservation*, London and New York: Routledge.

Joshi, V. (undated), *Living with risk in the Indian Himalaya region*, [Online: Web] Accessed 11 July 2006,

URL: http://www.unisdr.org/eng/public_aware/world_camp/2003/english/Others/India.doc

Karan, Pradyumna P. (undated), *Himalayas*, [Online: Web], Accessed 22 Sep 2006, URL: http://www.uttaranchal.ws/him.htm

Khawas, Vimal (2009), "Environmental Challenges and Human Security in the Himalayas", in (Eds) Somayaji, Sakarama and Ganesha Somayaji, *Environmental Concerns and Sustainable Development: Some perspectives from India,* New Delhi: TERI, pp 32-81

Khawas, Vimal and Ram Kumar Tamang (2005), "Conservation and Management of Water Resource in Sikkim Himalaya: some suggestions", *Spatio-Economic Development Record*, 12 (4): 27-31, July-August.

Khoshoo, T.N. (1993), "Himalayan Biodiversity Conservation - An Overview", in U. Dhar, (ed.) *Himalayan Biodiversity: Conservation Strategies,* Nainital: G.B.Pant Institute of Himalayan Environment and Development.

NBSSLUP (2004), *Soil Resource Management Reports*, National Bureau of Soil Survey and Land Use Planning: Nagpur.

Rao, K. S., and K. G. Saxena (1994), *Sustainable development and rehabilitation of degraded village lands in Himalaya.* Himavikas Publication Number 8, Dehradun: Bishen Singh Mahendra Pal Singh.

Rodgers, W.A. (1985), "Biogeography and Protected Area Planning in India", In J. Thorsell (ed.) *Conserving Asia's Natural Heritage*, Glad: IUCN.

Schroeder, Robert F. (1985), "Himalayan Subsistence Systems: Indigenous Agriculture in Rural Nepal", *Mountain Research and Development*, 5 (1): 31-44.

Schweinfurth, U. (1957), "Die horizontale und vertikale Verbreitung der Vegetation im Himalaya", *Bonner Geogr. Abh.* 20, Bonn.

Sharma, P. D. (2004), "Managing Natural Resources in the Himalaya", *Journal of the Indian Society of Soil Science*, 52 (4): 314-331.

Singh, R. L. (ed) (1971), *India: A Regional Geography*, Varanasi: National Geographical Society of India, reprinted 1991.

Singh, Tej Vir and Jagdish Kaur (eds) (1989), *Studies in Himalayan Ecology and Development Strategies*, New Delhi: Himalayan Books.

Tejwani, K. G. (1984), "Watershed Management in the Himalayas", Symposium on Effects of Forest Land Use on Erosion and Slope Stability: Hawaii.

Valdiya, K. S. (1998), *Dynamic Himalaya*, Hyderabad: Education Monographs University Press.

Vohra, C. P. (1981), "The Climate of the Himalayas", in J.S. Lall, (ed) *The Himalaya: Aspects of Change*, New Delhi: Oxford University Press.

Wadia, D. N. (1966), *The Geology of India*, MacMillan.

Yonzon, G. S. (2005), "Importance of the Himalayan Environment and its Impact on the Indian Subcontinent", *Tibetan Bulletin*, 9(2), [Online: web] Accessed 9 August 2006, URL http://www.tibet.net.

2

Environmental Degradation, Climate Change and Disasters in the Eastern Himalayas

Walter Fernandes

Climate change has become a concern in the Himalayas in general and in the Northeast in particular. The Eastern Himalayas experience the impact of environmental degradation in the form of floods, droughts and landslides. They result from the development policies that treat forests and other natural resources only as a commodity. The fragile ecology of the Himalayas that are "young" hills cannot bear the impact of the extent of human interference that it involves. But the policy makers continue to treat the natural resources only as a raw material to produce consumer goods for the middle class and for the profit of the investor and ignore the floods, droughts, landslides and impoverishment that this approach causes. This paper will reflect on that approach and study its impact.

Disasters: Natural or Human-Made?

This impact itself is not new. The result of deforestation

and other interference with the natural processes began to be felt already in the 1970s. A major alert was sounded in the 1980s when the Moore Island was born in the Bay of Bengal because of soil erosion caused by deforestation. Scholars like Prof. Madhav Gadgil calculated that in those days the country lost 5,334 million tonnes of soil per year because of deforestation. 29 percent of it was "exported" to the Bay of Bengal. The Moore Island that resulted from it became the centre of an international Indo-Bangladesh dispute (Gadgil et al. 1983: 128-129).

Soil that was not washed out to the sea remained in India. The riverbed rose because of it. According to the Centre for Science and Environment (1987), by 1980 the level of the Ganga had risen by 5 metres and that of the Brahmaputra by 1.98 m. Data compiled in the mid-1980s by the Centre for Policy Research, New Delhi (CPR 1985), showed that because of deforestation in their catchment area the silting of major dams was 3 to 4 times higher than what was foreseen at their planning stage. That reduced their lifespan substantially.

The situation continues to be grim as the frequency and intensity of floods show. Floods too have been part of life in India but they have got intensified in recent decades because of environmental degradation. A study done by K. Prasad (2005) of the Institute for Resource Management and Economic Development, New Delhi shows that in the 1950s, the biggest number killed by floods in India as a whole was 865 in 1955. The average number per year was 375. That began to change already from the 1960s. More than 1,000 persons died a year during 3 years of that decade and the average per year rose to 887. By the 1990s the average had gone beyond 15,000 per year and now it is more than 20,000. Studies also show that most floods and related disasters are human-made, caused by deforestation in the catchment area of the rivers and by the rising of the riverbed.

Disasters in the Northeast

The Northeast has been feeling the impact of environmental degradation already from the 1980s. Assam has been experiencing floods more than many other states do because the Brahmaputra is the biggest river of India and its level is rising as a result of deforestation in its catchment area. The impact is being felt from the 1980s while much of India felt it earlier because of what Madhav Gadgil (1989) calls sequential exhaustion of the forest and other resources. Deforestation began in the 19th century in western and southern India and moved north and eastwards from there. In the 1980s after denuding the forests of Jharkhand and Orissa began the attack on those of Uttaranchal and the Northeast. Even paper mills in South India got bamboo from Arunachal Pradesh, Assam and Mizoram (Meher-Homji 1988).

Floods are a result siltation and rise in the level of the riverbed. Floods were an annual event in Assam too but today they have become more frequent and intense. For example, in 1953 they affected 410,000 persons in Assam. In 1998 their number had risen to 4,698,000. In 2004 the worst floods in 50 years killed many hundreds of persons (Patowary 2004). Flash floods in its Goalpara district on October 9, 2004, caused allegedly by Bhutan releasing water from a dam, killed around 200 persons. In 2007 floods displaced some 4,000,000 persons and affected 6000,000 more since they came in 3 stages (Bhattacharjee 2004).

That is the difference between floods today and in the past. Till the 1960s they used to come in one stage in July and were part of the ecology and culture of the state as one can see from Bihu, the harvest festival. Though Assam and the rest of the Northeast depend mainly on the South Western monsoons, they celebrate the harvest festival not in September but in January as regions depending on the North-

Eastern monsoons do. It is because traditionally floods came in the early part of the monsoons and washed away most of the crops but also brought silt that functioned as soil nutrient. After the floods the farmers harvested the paddy that survived, replanted the rest and reaped it in December since the region got also the North Eastern monsoons. The floods caused some destruction but also brought nutrients with them. The farmers thus built their culture around this ecological event (Bharali 2007).

That has changed with their intensification. Floods have now become destructive. Till the 1980s they came in July. Today they come in three stages, in July, August and September and even in October. Moreover, the soil that is washed away does not function as a nutrient any more. Studies indicate that some seven percent of the land in Assam has been lost permanently to erosion. Floods displace many more persons than in the past. The affected persons do not have reserves in the form of seeds for a second crop or food for the lean season since frequent floods impoverish them and force them to consume even the seeds. Silt that covers much of their land renders it infertile since much of it is sand (Fernandes 2006).

As a result, even those who grow a second crop do not have adequate food for the rest of the year. Their frequency and intensity turn them into a human-made disaster and into a new problem without a solution (Bandyopadhyaya 2007). The traditional coping mechanisms that depended on a different frequency and intensity of the hazards cease to be viable. But their victims are unable to develop new techniques to cope with the changes. As a result, are impoverished, their economic and health status deteriorates. Social discrimination follows. In the absence of adequate resources to cope with the destruction for sheer survival the affected persons begin to overuse and destroy the forest and other resources that they had preserved as renewable till then. That begins the vicious circle of greater environmental degradation, more poverty and more destruction of resources (Fernandes and Bharali 2011: 305-307).

Search for Solutions

A solution has to be found to this vicious circle. The official solution is building of bunds to protect towns and cities. Ignored is the fact that the water has to flow somewhere. It moves away from the cities and causes floods in the remote rural areas. For example, bunds built to protect Dibrugarh, Jorhat and other towns in Upper Assam protect these towns but divert water to Majuli the biggest river island in the world. According to some studies, erosion has already destroyed more than 10 percent of it. Erosion destroys more land elsewhere and causes greater poverty particularly among the small and marginal farmers who abandon the little land they have and fill the urban slums (Goswami 2003).

The contractors and politicians gain from the floods since the central government supplies additional financial resources to the State for relief and to build bunds. That turns floods into a vested interest but the subalterns pay its price. Moreover, since with the disappearance of forests the soil does not absorb water any more droughts have become more frequent and add to the impoverishment caused by floods. Thus, the development policies have resulted in environmental degradation, floods and droughts. A solution to them requires a new development paradigm that respects nature and the communities depending on it (Dasgupta 2007).

Another solution is awareness of climate change that interference with nature has caused. In reaction to it some suggest that the Northeast be turned into a carbon sink for Europe (Fernandes 2009). On the surface this solution looks good but it ignores the fact that the Northeast is No 6 among the world's 25 mega-biodiversity zones. Biodiversity is people's livelihood and identity in this region. But a carbon sink needs greenery, not necessarily biodiversity. As a result, turning the region into a carbon sink for Europe may involve the Northeast becoming a region of commercial forests. That can destroy the people.

Conclusion

This quick analysis of the situation of North Eastern India shows the need to search for alternatives to the destructive measures that have resulted in environmental degradation. As a result the region feels the impact of climate change more than the rest of India does. One has, therefore, to search for solutions that can combine people's needs with environmental regeneration. Disaster coping mechanism such as community forestry may be the answer.

References

Bandyopadhyay, Chandrani. 2007. *Disaster Preparedness for natural hazards: Current Status in India.* Kathmandu: International Centre for Integrated Mountain Development.

Bharali, Gita. 2007. "Prakrotik Durjyoton Kiman Prakrotik?" *Dainik Janambhoomi,* September 8.

Bhattacharjee, Ratan. 2004. "Flash Floods Claim 52 Lives in Goalpara," *The Assam Tribune,* October 10.

CPR. 1985. *People's Participation as a Key to Himalayan Eco-System Development.* New Delhi: Centre for Policy Research.

CSE. 1987. *The Wrath of Nature: The Impact of Floods and Droughts on Floods and Droughts.* New Delhi: Centre for Science and Environment.

Dasgupta, Rajdeep. 2007. *Disaster Management and Rehabilitation.* New Delhi: Mittal Publications.

Fernandes, Walter. 2006. "Environment: For Whose Development?" *Jnanadeepa Journal,* 9 (n. 1, January), pp. 139-156.

Fernandes, Walter. 2009. "Climate Justice for the Northeast," *The Assam Tribune,* December 5.

Fernandes, Walter and Gita Bharali. 2011. *Uprooted for Whose Benefit? Development-Induced Displacement in Assam 1947-2000.* Guwahati: North Eastern Social Research Centre.

Gadgil, Madhav. 1989. "Forest Management, Deforestation and People's Impoverishment," *Social Action* 39 (n. 4, October-December), pp. 357-383.

Gadgil, Madhav, S. Narendra Prasad and Rauf Ali. 1983. "Forest Management and Forest Policy in India: A Critical Review," *Social Action* 33 (n. 2, April-June), pp. 127-155.

Goswami, D.C. 2003. "Flood Problem of Assam: Causes and Remedies," in Prabhat Chandra Sabhapandit (ed). *Flood Problem of Assam: Cause and Remedies.* New Delhi: Omsons Publications, pp. 8-14.

Meher-Homji, V.M. 1988. "Deforestation, Ecological Destabilisation and Drought," *Social Action* 38 (n. 2, April-June), pp. 115-128.

Patowary, Ajit. 2004. "Floods Inflict 990 crore Loss on Farm Sector," *The Assam Tribune,* Sept. 27.

Prasad, K. 2005. *Manual on Community Approach to Flood Management in India.* New Delhi: Institute for Resource Management and Economic Development.

3

Vulnerability Assessment of Rangeland Dependent Mountain Communities under Climate Change in the Hindu Kush-Himalayas (HKH)

Jasra A.W., S. Joshi & M. Ismail

Introduction

Rangelands are the largest land use in the Hindu Kush-Himalayan (HKH) region and provide enormous ecosystem services for people living upstream and downstream as well. These diverse and heterogeneous ecosystems are strategic in nature being backbone of the pastoral livelihood. Along with livestock production, these rangelands are hotspots and habitats of mountain biodiversity, vast watersheds with integrated interface while play important role in ecological buffering with other ecosystems like forests, wetlands etc. thus attracting considerable ecotourism. Despite high importance of rangeland resource, it has relatively obtained a little place in national, regional and international research and development agenda.

Climate change is being discussed as one of the most important drivers of change in the HKH region. Although it may be having both direct and indirect impacts on the socio-ecological systems at spatial and temporal scales (ICIMOD 2009), the understanding of this phenomenon is superficial. Studies from the perspective of pastoral communities indicate that the mountain pastoralists are exposed to highly vulnerable and risky situation (Angassa and Oba 2008, Rees *et al.* 2008). Eventually, livestock production has become more uncertain due to shrinkage of pastures, rapid decline in forage availability and changes in weather patterns (Thompson *et al.* 2009). The long term impact of climate change on the heterogeneous rangeland ecosystems in the HKH would be much more pronounced under rising temperatures and decreasing rainfall, the most important predictions being made by climate change models in many dryland areas (Dube and Pickup 2001, Retzer 2006). The climate change scenarios may stretch further the existing marginal resilience of the agro-pastoral communities beyond their adaptive capacity because of regular and frequent occurrence of droughts (Hailegiorgis *et al.* 2010). Such scenarios if not understood well in time, are more likely to cause rapid degradation of rangelands biodiversity, traditional knowledge pertaining to pastoral practices and erosion of HKH cultural heritage.

Klein *et al.* (2007) reported that warming induce loss of palatable plant species and increase in the production of non-palatable species. Consequently, climate change may potentially have negative impact on rangeland ecosystem goods and services. Conclusively measuring the nature and extent of socio-ecological vulnerabilities in relation to the adaptive capacity of rangeland dependent mountain communities remains critical task. Thus, this study was conducted to describe prevailing socio-economic conditions of the HKH mountain communities in the context of their dependence on rangeland resources and to assess their vulnerability under the climate change scenario.

Picture 1. **Rangelands under climate change in Ladakh**

Material and Methods

The study area

The study was conducted in six regional countries of the HKH region (Map 1) i.e. Afghanistan (Badakshan), Bhutan (Soi Yaksa and Nubri villages), China (Hongyuan County), India (Ladakh), Nepal (Upper Mustang) and Pakistan (Chitral). The six study sites were selected to represent the socio-economic conditions and climate change vulnerability of rangeland resources dependent mountain communities. The brief description of the study areas is summarized in Table 1.

Table 1
Basic description of the six study areas

Study area	*Overview of the study area*
Badakhshan- Afghanistan	Badakhshan is one of the provinces of Afghanistan, consisting of 29 districts. It is located in the north east of the country, between the Hindu Kush and the Amu Darya.
Soi Yaska and Nubri -Bhutan	Soi Yaksa is located at a distance of about three days walk from the nearest motorable road. The altitude ranges from 3900 msl to about 5000 msl. Whereas the winter camp of Nubri is located at 4200 msl compared to its summer station i.e. at 5000msl.
Hongyuan county- China	Hongyuan country covers 8350 km^2 with an average altitude of 3507 meter. The average annual temperature ranges from -36°C to 25°C and average annual rainfall is 728mm.
Ladakh –India	The Ladakh region, 32° to 36° N and 75° to 80° E of the Indian Trans-Himalaya, is a high altitude cold desert spread over approximately 45,000 km^2. The region is semi-arid to arid, with winter temperatures dropping below -25°C.
Upper Mustang- Nepal	Upper Mustang is bordering the Tibetan Autonomous Region (TAR) of China to the north and consists of 7 Village Development Committees (VDCs). The entire landscape of Upper Mustang is a cold desert with an altitude ranging between 3,300 and 6,480 m asl.
Momi (Chitral)-Pakistan	Momi community is located at a distance of 36 kilometer from the main Chitral town with an average altitude of 4000m. Annual temperature ranges between -14°C to 42°C.

Map 1. **Rangeland map of HKH region**

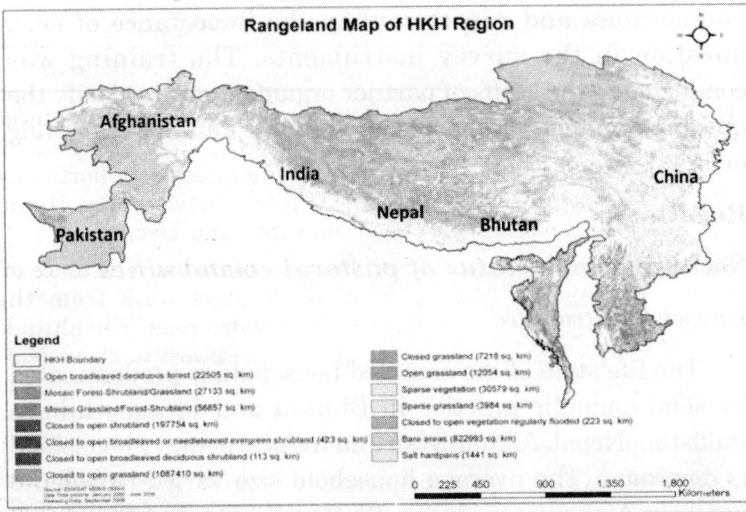

Data Collection

Household surveys of total 667 households were conducted in all six countries by using structured questionnaires. In order to have a representative sample, households in different settlements/villages of each study site were stratified into different strata depending on the degree of socio- ecological heterogeneities like nomadic, semi-nomadic or sedentary. The households were randomly selected for interview using the households list prepared for each stratum except Bhutan, where given the small population size of study areas, all households residing in both study sites were covered using the census survey. The total sample size of households was about 5% of the total households residing in the study area. A separate manual for field supervisors was prepared as a reference guide to conduct the household survey and focused discussions to ensure reliability and uniformity of collected data and information.

During March 25 to July 29, 2007, training programs and orientation sessions were organized in each of six countries to enhance the perception of the survey enumerators and supervisors on the importance of each question in the survey instruments. The training was conducted to the staffs of partner organizations to clarify the questions and the various codes in order to minimize sampling errors.

Results

Socio-economic status of pastoral communities

Household structure

The life style of the sampled household is predominated by semi-nomadic practice in Bhutan and China while in Pakistan, Nepal, Afghanistan and India sedentary household is dominant. The average household size varied between 6 heads in Afghanistan, India, Bhutan, China and Nepal to 8 heads in Pakistan (Figure 1). Majority of the sampled households were medium sized family whereas 34 percent households in India were small sized family.

Figure 1. **Average household size in the study sites**

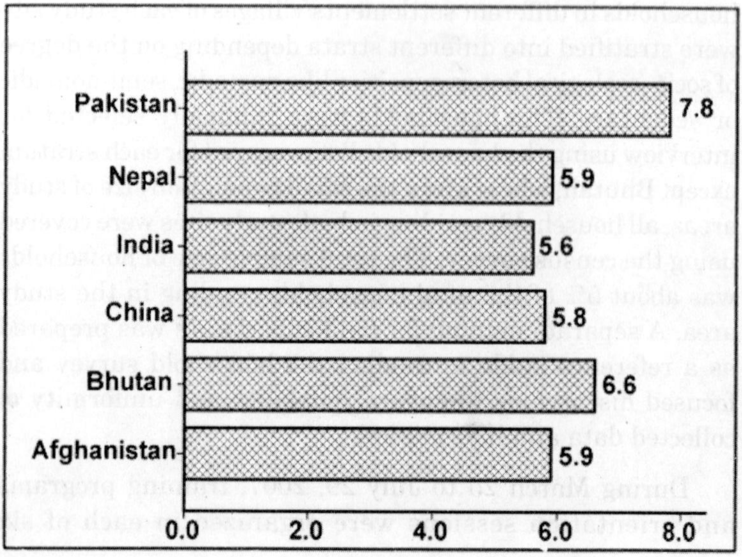

Literacy and occupational status

Education level of sample population was classified into seven categories; illiterate, no formal education, primary (1-5), lower secondary (6-8), secondary (9-10), higher secondary and above (11+). Literacy rate of households was highest in China (62%) followed by India (57%) and lowest in Bhutan (33%). The proportion of school age children not attending school was highest in Bhutan (43%) followed by Afghanistan (23%). Non-attendants were mostly girls (47%) especially in Bhutan whereas the opposite prevailed in other study sites. In most cases (except China) children were not attending school to assist parents in household activities.

Occupational status was classified based on questions concerning job titles and actual tasks. Majority of the sample population reported animal husbandry and agriculture as their main occupation. Majority population of households reporting employment as their main occupation was lowest in Afghanistan (1%) and highest in India (10%).

Water availability

River, stream and spring were the main sources of drinking water for the majority of the sampled households in all study sites. About 72% of the households in Afghanistan reported having access to spring water (Table 2). About 42 % of the respondents in Bhutan reported access to tap drinking water within house. About 66% of household in China reported access to rivers and streams. The quality of water from river and stream is generally perceived to be bad across all study sites except Bhutan. Quality of water of pond/well was perceived to be very bad in Nepal. The perception about adequacy of drinking water showed that spring water was less available in India and Nepal. One of the most important indicators of household access to water supply is distance to the water point. The average distance to fetch drinking water did not vary much across all study sites except in India where on average 11 km walk was required to fetch water.

Table 2

Percentage of households reporting sources of drinking water

Sources	*Bhutan*	*China*	*India*	*Nepal*	*Pakistan*	*Afghanistan*	*Total*
Spring	20.6	3.5	38.7	45.9	39.1	71.29	39.3
Pond/Well	0.0	30.6	0.6	1.7	19.1	3.96	9.27
River/Stream	34.9	63.5	35.3	49.4	36.3	21.78	39.8
Tap water within House	25.4	1.2	0.0	0.0	1.4	0	2.47
Public tap outside Premise	19.1	1.2	12.1	12	2.3	2.97	5.44
Other(specify)	0.0	0.0	13.3	17	1.9	0	3.71
Total	100	100	100	100	100	100	100

Land availability and utilization

In China, sample population represented a typical pastoral community where farming is unfeasible hence land was used solely for grazing animals. The proportion of landless households in India was 52% while 100% of household in China are landless. Food crops dominated the cropping calendar of households in all study sites expect Bhutan where cash crops mainly vegetable were predominant. Total area cropped per households was 6.35 ha in Afghanistan compared to only 0.14 ha in India. Over 90% of the sampled HHs in most of the study sites except India did not produce sufficient food to meet the annual subsistence needs from their own production (Figure 2). Pastoral households have adopted unique strategies to cope with food deficit problem. Consumption of livestock products and purchase of other food items from livestock income, trade and seasonal outmigration for alternative employment are some common responses to food security.

Figure 2. **Percentage of food sufficient households**

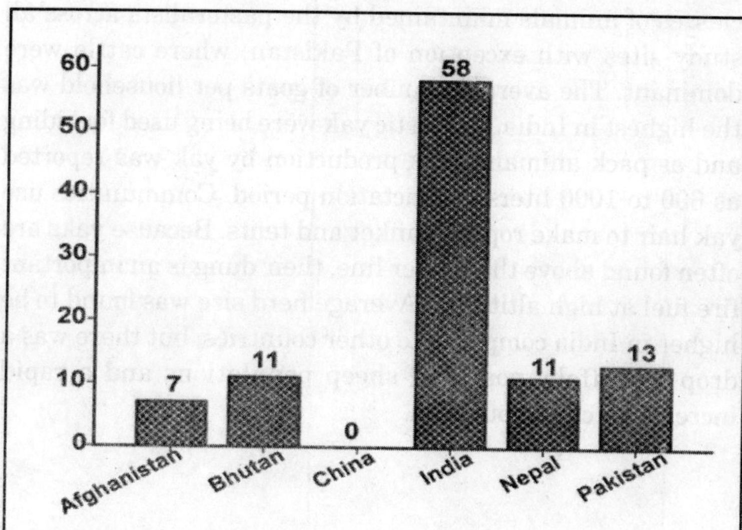

Infrastructure facilities

Physical inaccessibility and isolation continue to be a major bottleneck severely limiting the scope for improving the pastoral livelihoods in most study sites. Physical inaccessibility can be judged from average time taken to reach some major infrastructure/facilities in the study sites. About 80% respondents in Afghanistan and 62% in India reached the nearest facilities on feet. In Nepal, respondents had no other means of transport except their own feet to reach even the nearest destination. People residing in the study sites in Bhutan have to spend approximately 17- 20 hours to reach even the most basic social infrastructure like school, local markets and health centers. The survey revealed that all study sites were deprived of college, hospital, market center, electricity grid, and bank.

Livestock production

Livestock production was noted as an important component of the livelihoods of rangelands dependent mountain communities. Yak, horse and mule were dominant classes of animals maintained by the pastoralists across all study sites with exception of Pakistan, where cattle were dominant. The average number of goats per household was the highest in India. Domestic yak were being used for riding and as pack animals. Milk production by yak was reported as 600 to 1000 liters per lactation period. Communities use yak hair to make ropes, blanket and tents. Because yaks are often found above the timber line, their dung is an important fire fuel at high altitudes. Average herd size was found to be higher in India compared to other countries, but there was a drop in buffalo, goat and sheep populations and a rapid increase in calf population.

Picture 2. **Livestock grazing a summer pasture in Ladakh**

Annual household income

Obtaining reliable data on household income is seldom an easy task, especially when people have a general tendency to understate their actual income. Therefore, the data on household income needs to be cautiously interpreted. Sample household were asked to state their last year's total income from different sources. Average annual household income as estimated from various sources ranged from US$ 402 in Afghanistan to US$ 2781 in India (Figure 3). The per capita annual income was estimated to be the highest in India (US$ 536) whereas it was lowest in Afghanistan ($78). Income from livestock and the farm constituted the major part of household income in most of the study sites. With the growth of tourism and globalization, the contribution of livestock and farm animal husbandry to the household income is declining in Nepal.

Figure 3. **Average household income (US$)**

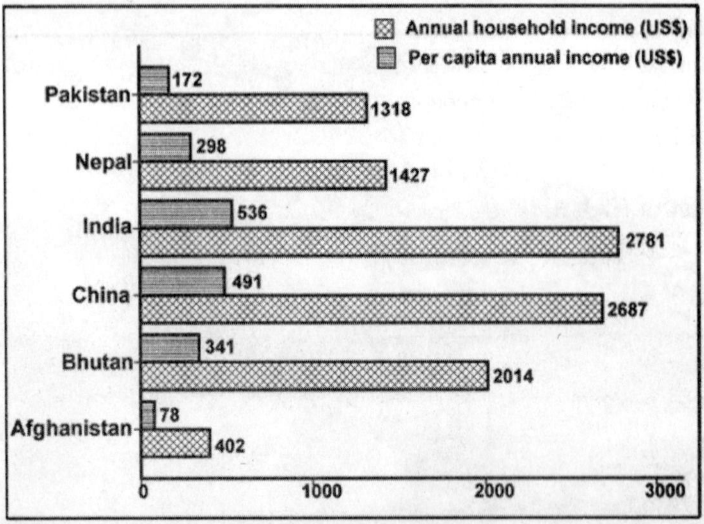

Climate change and potential vulnerabilities

The pastoralists are heavily dependent on natural resources, including water and forage [it means food and fodder], and are thus extremely vulnerable to climatic and other environmental changes. The livelihoods of HKH pastoralists depend to a large extent on weather, and pastoralist of the HKH were able to recall the pasture conditions and weather of at least the previous five years. Climate change impact was more pronounced because herders particularly women travelled longer distance to fetch potable water due to the drying up of water sources. On average, local people would walk 11km per day to fetch water for their daily needs in India. The accelerated rate of climate change has led to more difficulty in predicting rangeland productivity and changes in the availability of water and grass resources in most of the study sites, making the means and results of pastoral production more uncertain than ever before (Picture 3). Frequent droughts were reported as one of the major problems in most of the study sites. Pastoralist communities perceived that drought was occurring more frequently and over extended periods causing degradation of grazing lands. Changes in weather and climate were perceived as negatively affecting rangeland ecosystem, including changes in wildlife habitat, species composition etc. Prolonged summers and shorter winters were noted as indicators of climate change in most of the study areas.

Picture 3. **Rangeland degradation due to heavy grazing in Kailash**

Perceptions of pastoral livelihoods and vulnerability

The socio-economic vulnerability of pastoral communities was evaluated by asking respondents questions regarding the the existing situation of some major attributes of living conditions such as fuel, water, fodder, grazing area, number of livestock, etc. were getting better, the same as or deteriorating from earlier. Their response revealed that frequent droughts, shortage of fuel and grazing land, less water, food shortage, resource use conflict and poor market access were tome potential future crises across the study sites. An increasing trend of fuel shortage was a big concern of

Table 3

Household's perception towards changes in livelihood in percentage

	Crisis	*Food shortage*	*Water shortage*	*Fuel shortage*	*Fodder / grass shortage*	*Drought*	*Market in access*	*Loss of property*	*Loss of live-stock*	*Political crisis*
Bhutan	Increasing	3	3	65	42	43	18	3	11	56
	Decreasing	66	63	22	45	37	24	49	55	28
	Same as before	32	34	14	13	20	58	49	34	17
China	Increasing	6	75	18	46	61	21	47	73	100
	Decreasing	17	4	18	21	8	29	20	16	0
	Same as before	78	21	65	33	32	50	33	11	0
India	Increasing	14	45	37	28	20	28	12	64	5
	Decreasing	6	0	0	2	1	46	18	11	12
	Same as before	80	55	63	70	80	25	71	25	83
Nepal	Increasing	42	21	68	77	53	35	19	39	64
	Decreasing	35	46	17	10	9	51	21	22	0
	Same as before	23	33	15	13	38	14	60	39	36
Pakistan	Increasing	64	33	86	68	80	22	26	36	36
	Decreasing	21	25	12	19	14	15	14	41	10
	Same as before	16	41	2	13	6	63	59	24	54
Afghanistan	Increasing	59	67	62	61	49	38	24	54	0
	Decreasing	40	33	38	39	48	42	32	40	24
	Same as before	1	0	0	0	3	20	44	6	76

majority respondents in all the study sites except China and India where the fire fuel supply was unchanged (Table 3). Majority of the sampled households (64%) in Pakistan compared to 59 % in Afghanistan, and 42% in Nepal reported food shortage that was worse than before. Majority of respondents also agreed with the statement that women fell ill more often due to the smoky kitchen environment.

Discussion

The study provides several insights regarding the pastoral livelihoods and household perceptions towards vulnerability in the HKH region. Rangeland communities in HKH are amongst the poorest based on all indicators of their socioeconomic profiling. The average annual per capita income is estimated to range between US$78 in Afghanistan to $536 in India. There is hardly any evidence of livelihood diversification other than animal production in most sites with the exception of the Upper Mustang (Nepal), where tourism related business is emerging as an additional source of income. Forage and fuel shortage are commonly perceived as increasing trends although the problem is more pronounced in Nepal and Pakistan. Energy alternatives for mountain household are required to address this fire fuel shortage problem and to sustain grazing areas. Food deficits are an annual phenomenon and a persistent reality for a vast majority of sample households. Over 90% of the sampled households in most of the study sites do not have adequate food to meet the annual subsistence needs from their own production resources except India. This kind of insecurity would further encourage the downstream outmigration with additional social consequences in the HKH region.

Traditionally, rangeland people employ diverse coping strategies to overcome economic hardships. However, these traditional practices have weakened over time with globalization, modernization and climate change impacts.

Oba and Kaitira (2006) reported that knowledge of local resource dependent herders holds the potential to reduce uncertainty and increase the relevance of future assessments of vulnerability and adaptations to climate change. Physical inaccessibility and isolation continue to remain the major bottleneck that is severely limiting the scope for improving and diversifying the livelihoods in most rangeland study sites. Local development, particularly infrastructure, has its own implications for local culture and resource use patterns that need to be addressed carefully in the context of sustainable development for efficient and multiple resource use.

Based on the findings of this study, policy shifts in favor of rangeland resources are critical for livelihood diversification of pastoral communities in the HKH through sustainable development and multiple resource use by improving the rangeland ecosystem services, eventually enhancing local resilience as well as adaptive capacity. In the absence of alternative options, sustainable and optimum use of rangeland resources will remain a profound challenge (Adeel and Safreil 2008). The social consequences, in particular, may be severe for women who are at greater risk because of the nature of their household responsibilities. The socio-ecological dynamics of rangeland ecosystems in the HKH in the ongoing scenario need to be carefully understood to upgrade and integrate the community perceptions and indigenous practices with modern knowledge. Thus, the indigenous knowledge of the pastoralists in conjunction with the technical and scientific knowledge must be used to develop appropriate methods of resource management and livelihood diversification.

Inadequate feed supply is becoming the main constraint to livestock production. To overcome this problem, rangeland improvement and management must be directed towards key indicator plant species along with non-conventional feed resources to be introduced and piloted. For this purpose permanent ecological monitoring sites would be required to generate reliable data about developing early warning systems and forecasting for the pastoralists. In general, pastoralists in HKH are struggling very much to sustain their traditional way of securing livelihoods. Since the upstream rangeland ecosystems are equally vital to sustaining the downstream agricultural production systems by risk and hazard management, the mainstreaming of rangeland resources for optimum ecological management based on valuated ecosystem services is required.

[**Acknowledgements:** ICIMOD thanks Austrian Development Agency (ADA) for financial support of the study. Thanks are also due to all national partners from Afghanistan, Bhutan, China, India, Nepal and Pakistan for their help during the field surveys].

References

Adeel, Z and Safriel, U (2008) Achieving sustainability by introducing alternative livelihoods. *Sustain Sci* 3:125–133.

Angassa, A and Oba. G 2007) Herder perceptions on impacts of range enclosures, crop farming, fire ban and bush encroachment on rangelands of Borana, southern Ethiopia. *Human Ecology* 36: 201-215.

Dube, O.P and Pickup, G., (2001) Effects of rainfall variability and communal and semi-commercial grazing on land cover in southern African rangelands. *Climate Research* 17: 195–208.

Hailegiorgis A B., Kennedy, W G. Rouleau, M Bassett, J K. Coletti, M Balan, G C. and Gulden T (2010) An Agent Based Model of Climate Change and Conflict among Pastoralists in East Africa. *An International Environmental Modelling and Software Society (iEMSs) 2010 International Congress on Environmental Modelling and Software Modelling for Environment's Sake, Fifth Biennial Meeting, Ottawa, Canada. http://www.iemss.org/iemss2010/index.php?n=Main.Proceedings.*

ICIMOD (2009) Climate change and Hindu Kush-Himalayan rangelands, Information sheet 8/2009. Kathmandu: ICIMOD.

Klein, J.A., Harte, J and Zhao, X.Q (2007) Experimental warming not grazing, Decreases rangeland quality on the Tibetan plateau. *Ecological Applications* 17:541-557.

Oba, G. and Kaitira L.M. (2006) Herder knowledge of landscape assessments in arid rangelands in northern Tanzania. *Journal of Arid Environments* 66:168–186.

Rees, W.G., Stammler, F.M., Danks, F.S. and Vitebsky, P. (2008) Vulnerability of European reindeer husbandry to global change. *Climatic Change* 87: 199–217.

Retzer, V. (2006) Impacts of grazing and rainfall variability on the dynamics of a Sahelian rangeland revisited (Hein, 2006) – new insights from old data. *Journal of Arid Environments* 67: 157–164.

Thornton, P.K., van de Steeg, J., Notenbaert, A. and Herrero, M. (2009) The impacts of climate change on livestock and livestock systems in developing countries: a review of what we know and what we need to know. *Agricultural Systems* 101: 113-127.

4

Climate Change and Large Cardamom Farming in the Sikkim Himalaya

Vimal Khawas

Since the mid-1970s the average air temperature measured at 49 stations of the Himalayan region rose by 1°C with high elevation sites warming the most (Hasnain 2000, WWF 2005). This figure is twice as fast as the 0.6° C average warming for the mid-latitudinal northern hemisphere over the same time period (IPPC 2001, WWF 2005). Studies in Nepal (Shrestha *et al.* 1999) and Tibetan Plateau (Liu *et al.* 2002) indicate the rising temperature in recent times, with the warming being consistent and continuous after the mid-1970s in Nepal. The average warming in Nepal in its annual temperature between 1977 and 1994 was found to be 0.06° C per year. Incidentally, in both the countries the rate of warming is found to be more pronounced in the high altitude regions than the lower ones. Similarly, the lowland areas of India do not show significant warming trend (WWF 2005) indicating that the Himalaya is more sensitive and profoundly affected by climate change.

The question that rise in temperature will have a net positive, negative, or negligible impact on crop yields takes on a larger significance because additional hundreds of millions of people could be at risk of hunger or the window of opportunity (Cassman 2007). To answer this question, Lobell and Field (2007) used an innovative empirical/geostatistical approach to estimate the impact of increased temperature since 1980 on crop yields - a period when global mean temperature increased by 0.4æ% C. For three major crops — maize, wheat, and barley — there was a significant negative response to increased temperature. For all six crops evaluated (also including rice, soybean, and sorghum), the net impact of climate trends on yield since 1980 was negative (see Cassman 2007, Lobell and Field 2007). While the approach used by Lobell and Field can be questioned on several points, the body of their work represents an ambitious global assessment of recent climate impact on crop yields (Cassman 2007). Such scientific findings may be very relevant and informative for a largely agrarian economy like India. India is a second largest producer of wheat and the fifth largest producer of maize in the world. While the impact on yields may go unnoticed at a micro level the outcomes at global production may be substantial.

The third assessment report of the IPCC specifically considers agriculture as highly vulnerable to impacts of climate change in the South Asian region. In South Asia food security therefore becomes a primary concern. Crop production and aquaculture would be threatened by thermal and water stresses, sea-level rise, increased flooding, and strong winds associated with intense tropical cyclones. Acute water shortages combined with thermal stress should adversely affect wheat, and more severely, rice productivity in India given even a positive 'fertilisation' effect through an elevated atmospheric CO_2 level in the future.

Statement of the Problem

There has been a steady decline in the yield of large cardamom in Sikkim over the years. The yield used to be 250 kg per hectare in 1975-76 which came down to as low as 153 kg per hectare in 1996-97. One of the important factors of declining yield of the crop is attributed to old age of cardamom bushes. However, even more important factor in this regard has been the destruction of cardamom orchards by diseases in recent years. There are increasing indications from the scientists, policy makers and people on the field that change in temperature and rainfall pattern in the region over last many years could be the potential factors of destruction and declining productivity of large cardamom plantations in Sikkim.

Aim and objectives

The primary aim of this study is to examine and understand the relationship of climate change with the overall health of large cardamom plantation and rural economy in the Sikkim Himalaya. It has the following major objectives:

1. To inspect the trend of temperature and rainfall in Sikkim,
2. To study the production and yield of large cardamom in Sikkim,
3. To identify the major forces/factors that have influenced the health, production and yield of large cardamom,
4. To look into the role of climate change in the declining yield and extermination of large cardamom,
5. To identify the various [both adaptive and mitigative] measures/actions taken by the farmers and the government to tackle the declining productivity and destruction of large cardamom.

Methodology and Database

The study is carried out both at regional and local levels. Dzongu, one of the major cardamom producing areas of Sikkim, located in North Sikkim was selected for the purpose of fieldwork and primary information. The work is mainly based on published/unpublished reports of the Government of India and Sikkim, academic publications, NGO reports, reports in the regional news dailies, interaction with the policy makers/policy planners and grassroots workers, field observations and focused group discussion with the farmers. The study is based on empirical information and hence is largely descriptive in nature.

Climate Change in the Sikkim Himalaya

With the impacts of global warming becoming more apparent, Sikkim Himalaya is, perhaps, most in tune to the signs of change brought about by climate warming. The people across the towns and villages of Sikkim narrate revealing insights on how global warming is affecting their lives and livelihood. Meteorogical Department records reveal that between 1958 and 2005, there had been a slight change in the climate of Gangtok. According to Dr K Seetharam, Director of the Meteorology Centre, Gangtok[1] maximum temperature has been rising by 0.2° C per decade and minimum temperature has been falling by 0.3° C per decade. The annual rainfall has been increasing by 49.6mm per decade.

Further, officials at the Gangtok Meteorological office inform that for the past three years, the temperature in the state capital - located at an altitude of 5,480 ft - have always been on a higher side with the nights becoming warmer with each passing winter. The impact of global warming on the hill town is, further, evident from the temperature recordings of 30 years - from 1957 - 97. The average minimum

[1] Reported by Dionne Bunsha, *Frontline*, July 5 - 18, 2008.

temperature for January was 4 degrees Celsius for this period while the maximum temperature was 12 degrees Celsius. From 1997 onwards, the maximum temperature for January has been up by three degrees and the minimum by four.[2]

The state has also been witnessing prolonged periods of dry spell during winters often triggering rampant bush fires. In 2006-2007, forest fires were witnessed all over the state from January to March, resulting in an extensive loss of flora and fauna.[3] Bush fires did not spare even the higher altitudes of Lachung and Lachen (both above 8,000 ft) in North Sikkim and the army had to be called in. In short, less moisture is entering Sikkim during winter but there has been a lot of evaporation. Sikkim experienced its longest ever recorded, seven month long, dry and warm winter during 2008-09. The physical effects of it included record-breaking forest fires incidents, failure of winter crops and forced changes in the timing of subsequent crops.[4]

Higher temperatures have brought new crops to the mountains. Most of the seasonal flowers, which were supposed to bloom between March end and early April, are already blooming before February end. There are marked evidences of various birds, animals and insects changing their habitation. Amazingly, they are adapting to their new habitations and becoming habitual in the high altitude regions. Moreover, not only are the animals, birds and insects are changing their habits and habitats, the tree line is also displaying signs of global warming by shifting upward.

There are copious evidences of the low altitude tree species that are gradually shifting towards high altitude zone as the climate there are becoming salubrious enough for them to

---

[2] Reported by *The Telegraph* Wednesday, January 9 , 2008.

[3] Reported by *The Telegraph* Wednesday , January 9 , 2008.

[4] Reported by Rai subash and Parvinder Kaur, 'Get Rational about Climate Change not just Fashionable', *Now*, April 12, 2009.

survive. *Uttis,* a hot climate species, is now abundantly found in places above Gangtok. If the low altitude species are replacing the high altitude species, naturally the latter will reach the state of extinction. Further, the low altitude trees are being replaced by shrubs like *banmara* and others kinds of weeds known to destroy the forests.

It has also been found that Glaciers in Sikkim Himalaya are not behaving normally in recent years. For instance, the Jemu Glacier [located in North Sikkim] retreated by around 20 meters per year during 1975-1990 (Bahadur 2004: 53). Small streams that feed the large rivers are drying up more recently in and around the major watersheds of the region. This has not only affected the volume of the major rivers but also impacted the delicate relationship of flora and fauna and human habitation, particularly the livelihood of the poor rural hill folks.

Keeping in mind the worldwide debate on global warming and its likely impact on the glaciology and other natural resource base of the state, the Government of Sikkim recently appointed a high level national expert Group/ Commission headed by glaciologist Prof Syed Iqbal Hasnain. The Commission is expected to examine and report all the significant issues related to glaciers in and around Sikkim Himalaya.

> 'Sikkim is a mini theatre which in a way displays how climate change triggered by non-natural forces at the global level could bring disastrous natural calamities. We are worried in Sikkim as we have seen warm winters, increased flooding, landslides and rock avalanches from destabilised slopes. The torrential rains and unusually prolonged monsoon in 2007 caused extensive damage due to landslides. This was definitely not the case when I was young', *Chief Minister of Sikkim, Dr Pawan Kumar Chamling, while inaugurating the Commission's first meeting in January 2008.*

Agriculture in the Sikkim Himalaya

Sikkim is primarily a rural and agricultural economy where over 60 per cent of its population is directly engaged in agriculture. Principal food crops cultured in the region include maize, paddy, barley, millet, wheat, buck wheat, beans etc. Important cash crops are cardamom, ginger, potato, soybean, fruit crops and vegetables etc.

Horticulture is one of the major economic activities of the people of Sikkim. Large cardamom, ginger and turmeric are the principal crops while Mandarin orange, guava, mango, banana etc. are the principal fruits grown in the state. Sikkim is also a paradise for flowers. Gladioli, anthuriums, lilliums, primulas, rhododendrons, orchids as well as many other floral species thrive here. The state is home to an amazing 450 species of exotic orchids. There is immense potential for developing floriculture on a commercial basis here. The Government of Sikkim is making concerted efforts to turn this sector into an export-oriented industry.

Large Cardamom: A Major Cash Crop

Large cardamom (*Amomum sublatum Roxb*) locally known as "Allainchi" is the most important cash crop of Sikkim. It has been the main cash crop of Sikkim for generations and is an important contributor to the local economy. Large cardamom is planted between June and July when there is enough moisture in the soil. It is normally harvested in August. The crop grows well under the shade of forest trees at altitudes ranging from 1000-2000 metres with a rainfall of 3000-3500 mm per annum. Deep and well-drained soils with a loamy texture are best suited for cardamom (Pathak 2008). Compared to traditional crops like maize and millet, the income from cardamom is much higher and therefore facilitates to alleviate many smallholder households from poverty.

Over 16,949 farmers cultivate large cardamom in the Sikkim Himalaya of which 15, 209 are small and marginal farmers, while around 30 per cent are totally dependent on this crop for their survival. Large cardamom plantation covers about 26 thousand hectares of land under cultivation. With about 54 per cent of the global market for large cardamom, India is the largest producer in the world, followed by Nepal with 33 per cent of the market and Bhutan with 13 per cent.[5] Of the Indian output, almost 88 per cent used to come from Sikkim till about a decade ago, making it the largest producer of large cardamom in the world. However, it is believed, Sikkim today produces only about 30 to 40 per cent. Besides Sikkim, Darjeeling district of West Bengal and other Northeastern states of India like Mizoram, Manippur, Meghalaya, Nagaland and Arunachal also culture large cardamom.

Production of Large Cardamom within India

Region	*Per centage Share*	*Production (MT)*
Sikkim	87.5	3500
Darjeeling	12.5	500

Production of Large Cardamom in the World

Region	*Per centage Share*	*Production (MT)*
India	53.3	4000
Nepal	33.3	2500
Bhutan	13.3	1000

Source: For both Table 3.1 and 3.2 Berring et al. 1993 and Mande et al 1996.

[5] It is used as a spice in several ayurvedic preparations. It contains 2-3% essential oil and possesses medicinal properties. Large cardamom has a pleasant aromatic odour, due to which it is extensively used for flavouring vegetables and many food preparations in India. It is also used as an essential ingredient in mixed spices preparation. Apart from aroma, large cardamom also has high medicinal value. The decoction of seeds is used as a gargle in infection of teeth and gums. Large cardamom seeds are considered as an antidote to either snake venom or scorpion venom. It is also reported that large cardamom seeds are used as preventive as well as curative measure for throat troubles, congestion of lungs, inflammation of eyelids, digestive disorders and in the treatment of pulmonary tuberculosis.

Depletion and Destruction of Large Cardamom in Sikkim

There has been a steady decline in the yield of large cardamom in Sikkim over the years. The yield used to be 250 kg per hectare in 1975-76 which came down to as low as 153 kg per hectare in 1996-97. One of the important factors of declining yield of the crop is attributed to old age of cardamom bushes (Lama 2001, 2007). According to the officials of the Spice Board the crop has been ailing since the 1980s. The decline began with two successive and prolonged bouts of viral infections (chirkey and foorkey[6] virus) that hit large cardamom growing tracts in Sikkim, North Bengal, Nepal and Bhutan. Surveys during 2002 and 2003 in the Darjeeling-Sikkim Hill area found that as much as 27 per cent of plants were affected (Mandal *et al* 2004).

A disease caused by a soil-borne fungus, Fusarium oxysporum, also affected the already weakened crop. Though this fungus has always been present in the soil here, botanists declare that a mix of environmental conditions, like deforestation and acidification of soil with old plantations and poor management, allowed the fungus to spread, affecting more than 60 per cent of the crop. The diseases have ruined the farmers' livelihood and local as well as regional economy.

Pathak (2008) observes:

> Fungal or bacterial diseases are seldom reported in large cardamom. Only minor diseases like leaf streak or rot diseases are found in isolated areas. The major threat to large cardamom is the widespread occurrence of viral diseases, viz., chirkey and foorkey. These diseases are seen throughout the large cardamom growing tracts of Sikkim and Darjeeling and cause considerable crop loss. *These diseases have spread due to drastic change in the ecosystem, inadequate rain in dry months* **[emphasis added]** and absence of good agricultural practices by the

[6] Also known as Cardamom Bushy Dwarf Virus.

farmers. Many cardamom farmers failed to plant varieties suitable to their altitude.

Chirkey: The symptoms are characterized by mosaic appearance on the tender leaves with pale streaks, which slowly turns into brown, resulting in withering and drying of the plants. Growth and yield of the affected plants gradually declines and ultimately they perish. The disease is transmitted by aphids. It also spreads by planting infected suckers. Transporting of infected suckers from one area to another leads to the spread of this disease. The disease is also transmitted mechanically through the knife used for harvesting.

Foorkey: Numerous small tillers appear at the base of the affected plants which become stunted and fail to give any yield. Even the inflorescence is noticed to produce unproductive spikes.

Leaf eating Caterpillar: Initially the caterpillar of the moth Artona Chorista feeds on the leaf lamina from under the surface of the leaf and finally defoliates the leaf completely leaving only the midribs. Their incidence is noticed in May-July and October-March. At present, their natural enemies keep these insects under control. If insecticides are used to control them, then their natural enemies will also disappear which may lead to an outbreak of these pests in epidemic form. The best method of control is to inspect the plantations during May-July and October-March, to handpick the infected leaves along with the caterpillars and destroy them by burning.

District-wise area, production and yield of large Cardamom

District	*2004-05 Area ['000 Hectare]*	*2005-06 Pro-duction ['000 Tonnes]*	*Average Yield [per Hectare]*	*Area ['000 Hectare]*	*Pro-duction ['000 Tonnes]*	*Average Yield [per Hectare]*
North	8.99	1.38	154	6.73	0.98	145
East	6.77	1.00	148	5.80	1.03	177
South	4.52	0.66	146	3.84	0.61	160
West	4.53	0.70	155	3.60	0.54	150
State	24.80	3.74	151	19.97	3.16	158

Source: Sikkim: A Statistical Profile 2006-07, pp.110.

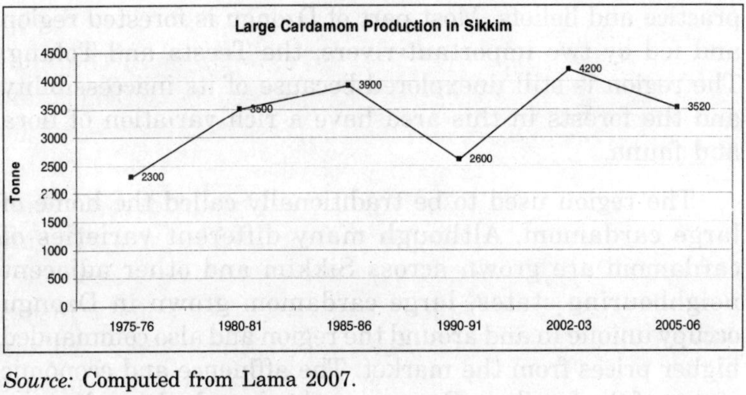

Source: Computed from Lama 2007.

Pathak (2008) suggested following measures to check the spread of viral diseases in cardamom plants:

i) Keep a constant vigil to detect disease affected parts.
ii) Uproot and destroy affected plants as soon as symptoms appear. Repeat detection and uprooting at regular intervals.

iii) Use seedlings produced in certified nurseries.

iv) Propagation through suckers is recommended only through certified multiplication nurseries.

A Case of Dzongu Valley in North Sikkim

Dzongu Valley is a place located in the northern part of Sikkim and is believed to be oldest inhabited region of the Lepchas. Not very far away from Mangan, the headquarters of North Sikkim District; Dzongu is still an untainted and least populated area of the Himalayan Sikkim. Earlier, Dzongu was restricted to the Lepchas, but in the recent years, it has been made accessible to the tourists. Lying at the close proximity to the Kangchendzonga Biosphere Reserve at an elevation ranging between 3000 ft and 20000 ft above sea level, Dzongu is an unpolluted and least trodden territory in Sikkim Himalaya. Still uninfluenced by modern civilisation, the traditional Lepcha Communities in Dzongu live in relative isolation from the outside world with their own traditional practice and beliefs. Most part of Dzongu is forested region and fed by two important rivers, the Teesta and Tolung. The region is still unexplored because of its inaccessibility and the forests in this area have a rich variation of flora and fauna.

The region used to be traditionally called the home of large cardamom. Although many different varieties of cardamom are grown across Sikkim and other adjacent neighbouring states, large cardamom grown in Dzongu occupy unique in and around the region and also commanded higher prices from the market. The affluence and economic status of the family in Dzongu used to largely depend on the production of large cardamom. All families traditionally inherited large cardamom farms from their fore fathers.

Demise of Large Cardamom in Dzongu

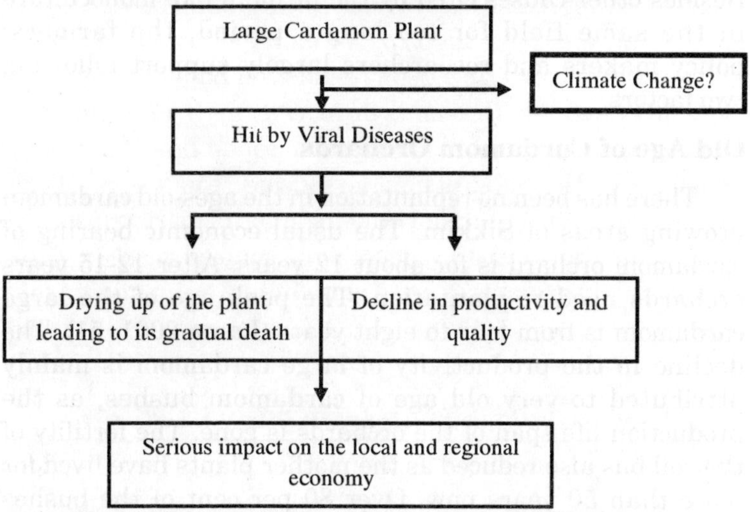

A typical household in Dzongu used to produce about 80-90 sacks of large cardamom and sell around 1000 kgs annually. Diseases called chirkey and foorkey have seriously attacked cardamom plants in the last eight to ten years. The plants gradually dry up, drastically reducing productivity and quality of cardamom seeds, and ultimately die. Leaf eating caterpillar and stem borer have also been reported to be present in cardamom plants in recent times. The production of large cardamom has gone down by more than 80 per cent in and around Dzongu region during the period. The quality of the cardamom seed has also drastically gone down. The cardamom seeds do not get fully ripped and matured during the harvesting period leading to low market price. People in the area had never anticipated such sorry state of the crop in the wildest of their dreams.

Major Causes Attributed to the Devastation of Large Cardamom

Several reasons have been attributed both by the farmers and the policy-makers regarding the increasing ill health and declining productivity of large cardamom orchards in Sikkim. Besides other causes cited by the farmers like monoculture in the same field for too long a period, the farmers, policy makers and researchers largely support following two factors.

Old Age of Cardamom Orchards

There has been no replantation in the ages-old cardamom growing areas of Sikkim. The usual economic bearing of cardamom orchard is for about 12 years. After 12-15 years orchards need re-plantation. The peak age of the large cardamom is from four to eight years (Lama 2001: 51). The decline in the productivity of large cardamom is mainly attributed to very old age of cardamom bushes, as the production life span of the orchards is gone. The fertility of the soil has also reduced as the mother plants have lived for more than 50 years now. Over 80 per cent of the bushes require immediate replantation. The absence of gap filling plants destroyed by chirkey, foorkey and other diseases and pests annually is equally detrimental.

Climate Change

There are no tangible scientific research findings on the ground as yet that validate the linkage of devastating blight epidemic on the cardamom crop to rising temperatures and fluctuating rainfall. However, the situation fits in, by and large, with the projections / predictions, on the plight of the agriculture due to climate change, made by the International Panel of Climate Change (IPCC) reports.

> 'In 1998-9, there was a very long spell of dryness in Sikkim. According to the estimates by the Agriculture Department, Government of Sikkim, the total rainfall was 93.34 per cent lower that that recorded over the last 25 years. This had visible adverse effect on the economy, especially on drinking water, rabi crops and cash crops (like cardamom, ginger and orange). Because of this prolonged drought farmers had to postpone the sowing of seeds and in many cases had to resort to not sowing the same. Farmers from the North district were the most affected. It is in

fact estimated that 60 per cent of the State's large cardamom has been lost in this unprecedented dry spell'. (Lama 2001: 52).

The blight outbreak was, however, reported after 2002. According to MR Sudarshan, deputy director of the Spice Board:

> 'There was a dry period from 2000-2002, which could have made the plants weak and could have resulted in the spread of the disease. It could be partly due to climate change. We don't have reliable data in all the cardamom growing areas – from low to high altitudes – to be able to make the connection. The spores of the fungus are carried through wet wind. The epidemic has become unmanageable in some places. The yield has fallen between 30 to 50 per cent'.

Scientists are, nonetheless, still investing the concrete reasons for the outbreak of diseases across the large cardamom plantations in Sikkim. They are struggling to scientifically prove the linkage of climate warming with increasingly dwindling health of the large cardamom in the state. However, large section of the farmers and policy makers believe that rise in temperature and variation in the monsoon and winter rainfalls are the major causes of cardamom's ill health. According to the farmers, cardamom orchards in the lower hills (*awul*) are severely affected by the diseases while those in the higher hills (*lekh*) are comparatively healthy even today.

Measures Taken by the Farmers

Adaptive Measures

Today the farmers have lost their hopes in large cardamom farming and have gradually resorting to diversify their economic activity. They are shifting to alternative farming practices such as cultivation of ginger, orange and other horticulture crops although the economic returns of these crops are relatively lower than that of large cardamom. In fact, these crops were traditional cultivated by the farmers at the subsistence level while large cardamom used to be the major income generator. With the gradual demise of the cardamom orchards, the farmers are now trying out crops like ginger and orange at commercial level to supplement the income that used to be generated by large cardamom. Horticulture crops like organic vegetables are also being tried out more recently. The farmers have also started increasing their livestock population.

Mitigative Measures

An important mitigative measure tried out by the farmers to save the dying cardamom plant was spraying of medicine on the cardamom orchards both at their own initiative and with the support of the government. However, spraying of medicine did not prove to be fruitful.

On the advice of the spice board and other government agencies, the farmers are also cleaning and burning the fields where cardamom orchards are totally destroyed. The government has advised the farmers to keep such fields fallow for some[7] years and to properly manure the cardamom field.

Further, replantation of the cardamom plants both on the weak/old orchard areas and virgin areas are being carried out by the farmers often with the help of the Spice Board and Agriculture/Horticulture Department of the state government.

Support from the Government

The Spices Board[8] has recently taken up various programmes to rejuvenate large cardamom plantations in the state through the following schemes:

---

[7] At least two years.

[8] Spices Board was constituted on 26th February 1986 under the Spices Board Act 1986 (No. 10 of 1986) with the merger of the erstwhile Cardamom Board (1968) and Spices Export Promotion Council (1960). Spices Board is one of the five Commodity Boards functioning under the Ministry of Commerce & Industry. It is an autonomous body responsible for the export promotion of the scheduled spices and production development of some of them such as Cardamom.

1) *Certified nursery scheme:* The Board supports the raising of nurseries in farmers' fields by offering a grant-in-aid of Rs. 10,000 per nursery producing 10,000 suckers.

2) *Replanting scheme:* This scheme is intended to encourage growers to take up replantation of senile and uneconomic gardens. A subsidy of Rs. 6,000 per hectare is offered to the growers.

3) *Supply of sprinkler irrigation units:* To tide over the drought situations during summer, the Board assists growers in procuring and installing sprinkler units and accessories by providing 50 per cent of the cost as subsidy with a ceiling of Rs. 2500 per set.

4) *Low cost driers:* In order to improve the quality of the cured large cardamom the Board helps the growers to replace their traditional *bhatti* curing system with improved driers by providing a subsidy of Rs. 10,000 per drier.

5) *Processing units:* The conventionally cured cardamom does not fetch the right price because of unscientific processing and packaging. In order to overcome this situation, the Board proposes to assist the growers in setting up their own processing/powdering/packaging units by providing assistance to the tune of 50% of the cost, subject to a maximum ceiling of Rs. 50,000 per unit.

Production of large cardamom has been, however, reducing in Sikkim despite the efforts of the Spices Board. About Rs 23 crore has been provided in the 11th Five Year Plan for the production of large cardamom in the state of Sikkim and in Darjeeling district of West Bengal.[9] For the

---

[9] Reported by *The Statesman,* February 17, 2009.

past two years, the Spices Board is concentrating on disease free planting material and replanting of cardamom bushes. About 1600 nurseries have been proposed for developing healthy seedlings to replace the old bushes in Sikkim. A subsidy of Rs 1.2 crore is expected to be disbursed among cardamom growers. The State Horticulture and Cash Crops Development Department is planning to replace the ailing cardamom plantations with organic vegetables or bamboo plantations.

Plate 1: **Relatively healthy Cardamom Plantation: Dzongu, North Sikkim**

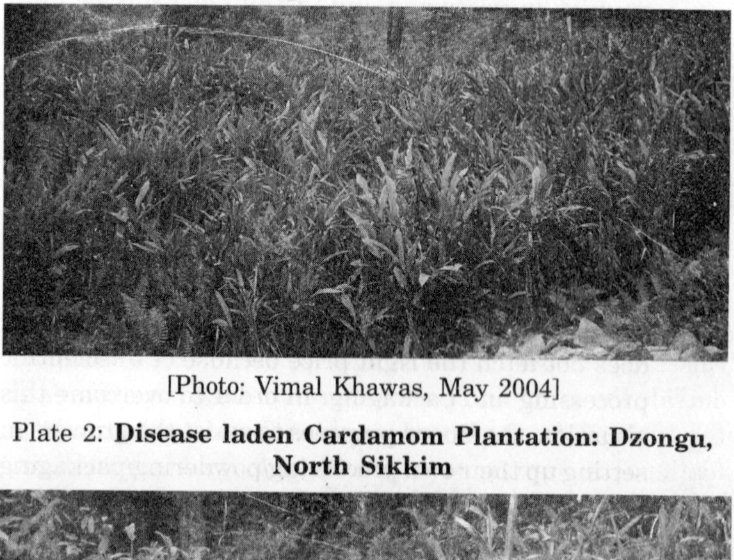

[Photo: Vimal Khawas, May 2004]

Plate 2: **Disease laden Cardamom Plantation: Dzongu, North Sikkim**

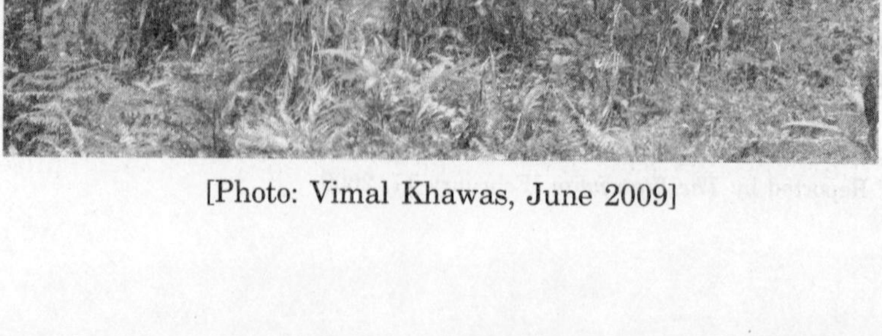

[Photo: Vimal Khawas, June 2009]

Conclusions

Sikkim Himalaya as an integral part of the Himalayan region has witnessed notable variation in its temperature and rainfall pattern over the years. According to Meteorological sources, maximum temperature in Gangtok has been rising by 0.2° C per decade and minimum temperature has been falling by 0.3°C per decade. The annual rainfall has been increasing by 49.6 mm per decade. The physical effects of such meteorological deviations include glacier retreat, GLOFs, record breaking forest fire incidents, failure of agricultural crops and forced changes in the timing of crops etc. Further, animals, birds and insects are changing their habits and habitats while the tree line is gradually shifting upward.

Sikkim is primarily a rural and agricultural economy where over 60 per cent of its population is directly engaged in agriculture. Principal food crops cultured in the region include maize, paddy, barley, millet, wheat, buck wheat, beans etc. Important cash crops are cardamom, ginger, potato, soybean, fruit crops and vegetables etc. Large cardamom is raised on commercial basis and is exported both within and outside the country. In fact, Sikkim has the largest area and the highest production of large cardamom in India. It is a foreign exchange earner crop of Sikkim. The state has limited industrial potential due to its geologic and geomorphic constraints.

Large cardamom farming has been suffering from decline in production and gradual drying and subsequent death of the plant in the last 1 to 1.5 decades. There has been a steady decline in the yield of large cardamom over the years. One of the important factors of declining yield is ascribed to old age of cardamom bushes. A section of the policy makers and farmers also believe that monoculture of cardamom plants in the same fields for generations, is perhaps, an added cause of declining productivity of the cardamom. However, a further more important and serious factor in this regard has been the destruction of cardamom orchards by viral diseases in recent years. There are increasing warnings from the scientists, policy-makers and people on field that change in temperature and rainfall pattern in the region over last many years could be the potential factors of the disease spreading viruses and subsequent destruction and declining productivity of large cardamom plantations across Sikkim.

Both adaptive and mitigative measures are being sought by the farmers to sustain their livelihood. These measures have been taken both at the household/community initiative as well as with the support of the government. In the first case, the farmers are resorting to alternative livelihood opportunities like experimenting other cash crops like ginger, orange, vegetables at commercial level. With regard to the mitigative measures, the farmers are spraying pesticides, cleaning the fields, keeping the land fallow for some years and then replanting the fields with fresh cardamom saplings. The government, especially Spice Board of India, has been helping the farmers since last many years to salvage their livelihood. The results of the mitigative measures have not been very fruitful, nonetheless.

To sum up, a fundamental question to be seriously taken here is: if Sikkim does away with its large cardamom farming in the near future what would be the fate of a large section of the agrarian population dependant on cardamom farming. It would not only affect the state's agricultural economy but also the Indian spice trade in a major way. Since, a large chunk people in the rural areas have been totally dependent on cardamom as the main cash crop providing them with a significant part of their livelihood, the impact of steadily falling health of cardamom orchards and declining yield is likely to be disastrous. Such a scenario calls for a need to workout and evolve a comprehensive strategy to revitalise the health of cardamom plants and sustain the yields. Moreover, plants affected by the viral diseases cannot be cured but the losses may be minimised by adopting appropriate management practices.

References

Assam Tribune. (2009). Gangtok getting warmer. March 28

Bahadur, Jagdish (2004). *Himalayan Snow and Glaciers: Associated Environmental Problems, Progress and Prospects*. New Delhi: Concept Publishing Company.

Bandyopadhyay, M.K. (1998). Glacier Variation in the Himalaya. *Geographical Review of India*, 60 (4): 381-391

Berrig Ch, Koechilin R, Tarnutzer A. (1993). *Sikkim: A Preliminary profile with particular emphasis on horticulture and animal husbandry*. Study Group on Institutions, Human Actions and Resource Management, Institute of Geography, University of Zurich, pp 47-51.

Bunsha, Dionne. (2008). Withering Heights. *Frontline*, 25 (14), July 05-18, [Online Web], Accessed: June 19, 2009, URL: http://hindu.com/thehindu/fline/f12514/stories/20080718251404300.htm

Cassman, Kenneth G. (2007). Climate Change, Biofuels, and Global Food Security. *Environ. Res. Lett.*, 2 (011002), pp. 1-3.

Fujita, K *et al.* (1998). Glaciological Observation of Yala Glacier in Langtang Valley, Nepal Himalayas: 1994 and 1996. *Bulletin of Glacier Research*, 16: 75-81

Fujita, K *et al.* (2001). Glaciological Observation of Rika Samba Glacier in Hidden Valley, Nepal Himalayas: 1998 and 1999. *Bulletin of Glaciological Research,* 18: 31-35

Government of Sikkim. (2003). *Sikkim- A Statistical Profile-2002*. Govt of Sikkim: Gangtok.

Hasnain, S.I. (2000). *Status of Glacier Research in the HKH region.* ICIMOD: Kathmandu.

Intergovernmental Panel on Climate Change. (2001). Climate Change 2001: The Scientific Basis. Contribution of Group I to the Third Assessment Report of the Intergovernmental Panel for Climate Change.

Intergovernmental Panel on Climate Change. (2007). *Climate Change 2007: The Physical Science Basis*. Summary for Policy Makers, WMO-UNEP- IPCC: Geneva.

Kadota, T *et al*. (1997). Monitoring and Prediction of Shrinkage of a Small Glacier in Nepal Himalaya. *Annals of Glaciology*, 24: 90-94.

Khawas, Vimal. (2008). *Environmental Issues and Human Security in the Himalayas: A Comparative Study of Eastern Nepal and Sikkim*. Unpublished PhD Thesis submitted to Jawaharlal Nehru University, New Delhi.

Khawas, Vimal. (2007). *Environmental Challenges and Human Security in the Himalaya*. Paper presented at National Seminar 'Himalayas in the New Millennium: Environment, Society, Economy and Polity, Centre for Himalayan Studies, North Bengal University, Darjeeling, January 11-12.

Khawas, Vimal. (2007). Global Warming and Climate Change: Implications for the Human Security in India. *Social Change*, 37 (3): 92-119.

Khawas, Vimal and Ram Kumar Tamang. (2005). Conservation and Management of Water Resource in Sikkim Himalaya: Some Suggestions. *Spatio-Economic Development Record*, 12 (4): 27-31, July-August.

Lama, M.P. (2001). *Human Development Report: Sikkim*. Gangtok: Government of Sikkim.

Lama, M.P. (2007). *Sikkim Economic Survey 2006-07*. Gangtok: Government of Sikkim.

Liu, Shiyin *et al*. (2002). Glacier Fluctuation and the Inferred Climatic Changes in the A'nyemaqen Mountains in the Source Area of the Yellow River, China. *Journal of Glaciology and Geocryology*, 24 (6): 701-707.

Lobell, D B and Field, C B. (2007). Global Scale Climate-Crop Yield Relationships and the Impacts of Recent Warming. *Environ. Research Letter,* 2(014002), pp. 1-7.

.Mandal, B.: S Mandal; K B Pun; ACPV Varma. 2004. First Report of the Association of a Nanovirus with Foorkey Disease of Large Cardamom in India. *Plant Disease*, 88 (428), [published online].

Mande Sanjay, Anil Kumar, Raman P. and Kishore VVN. (1996). *A Study of Cardamom Curing Chambers in Sikkim with*

Reference to Energy Efficiency and Technology Upgradation. Final Report submitted to Swiss Development Co-operation, Bangalore.

Mayweski P. and P.A. Jaschke. (1979). Himalayan and Trans-Himalayan Glacier Fluctuation Since A.D. 1812. *Arctic and Alpine Research*, 11(3): 267-287.

Pathak, Anjali. (2008). Cultivation of Large Cardamom in Sikkim. *Ishani: Tribal World of the North East India,* 2 (6), November

Seko, K *et al.* (1998). Changing Surface Features of Khumbu Glacier, Nepal Himalayas Revealed By SPOT Images. *Bulletin of Glacier Research*, 16:33-41.

Sethi, Nitin. (2007). Climate Change Damaging Crops. *The Times of India*, April 3.

Shrestha, A B. *et al.* (1999). Maximum Temperature Trends in the Himalaya and Its Vicinity: An Analysis Based on Temperature Records From Nepal For the Period of 1971-97. *Journal of Climate*, 12: 2775-2787.

Shrestha, A B *et al.* (2000). Precipitation Fluctuations in the Himalaya and Its Vicinity: An Analysis Based on Temperature Records From Nepal. *International Journal of Climate*, 20: 317-327.

Siegenthaler, U *et al.* (2005). Stable Carbon Cycle-Climate Relationship during the Late Pleistocene. *Science*, 310, pp. 1313-1317.

WWF .(2005). *An Overview of Glaciers, Glacier Retreat and Subsequent Impacts in Nepal, India and China*. World Wide Fund for Nature: Kathmandu.

Yamada, T. (1993). *Glacier Lakes and Outburst Floods in Nepal Himalayas*. Proc. Intern. Symp. on Snow and Glacier Hydrology, Kathmandu, Nepal (16-21 Nov. 1992), IAHS Publication. No. 218: 319-330

5

Environmental Constraints and Population Pressure in a Mountain Ecosystem of Western Himalaya: A Case Study of Kinnaur

Uttam Lal

Introduction

Kinnaur is a remote Himalayan district with altitude ranging from 1300m to above 6000m above mean sea level. It is situated in the eastern part of Himachal Pradesh. The district lies between 31°05'55"N and 32°05' 20" N latitude and between 77°45'E and 79°00' 50"E longitude with an area of 6401 square kilometers and had a population of 77007 persons in 2001. It is bounded on its Northern and Northwestern side by spurs of snowy mountains that separate it from Spiti and in the east by similar terrain, which is in Tibet. Shimla district of Himachal Pradesh and Uttarkashi district of Uttarakhand are situated to its south. The climate ranges from moist-temperate to cold-arid. The entire district is composed of a series of mountains, which are alternated with a succession of valleys. The Satlej River and its tributaries traverse these mountains. Kinnaur has three parallel ranges, namely, the Zanskar, the Great Himalayan and the Dhauladhar Ranges. The Zanskar Range cuts off Kinnaur from Nagari region of Tibet and the Great Himalayan Range lies to the south of Zanskar Range. The Dhauladhar Range occupies the southern boundary of Kinnaur, separating it from Uttarkashi district of Uttarakhand and Rohru tehsil of Shimla district.

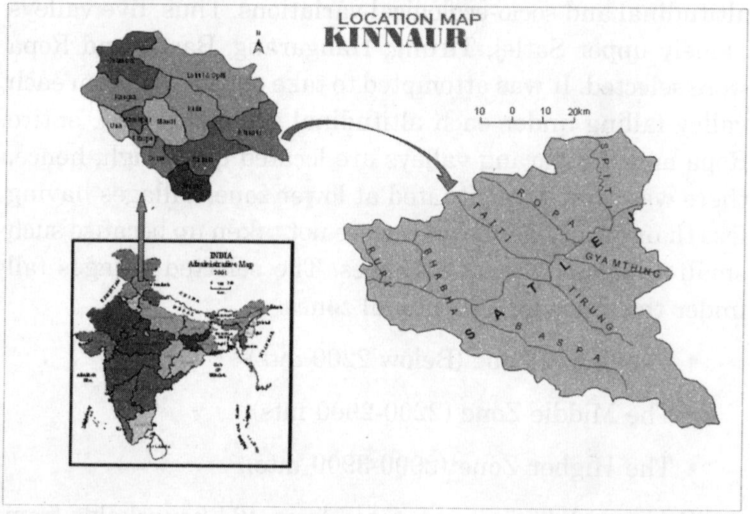

The Great Himalayan Range separates the moist-temperate region of the district from the cold-arid Trans-Himalayan region, which occupies a major area of Kinnaur district. On account of the high altitude and resultant cold-arid climate, it is among the most difficult regions to live in, environmentally. The impulse of development has been delayed in time due to its geographical remoteness, and therefore, the influence of the environment is pronounced. In response to environmental compulsions, Kinnaurese developed a pastoral-cum-agricultural economy to meet local food requirements.

Methodology of Data Collection

It involves two steps in the present study: selection of villages and selection of households.

Physiographic and climatic variations are significant to understand population and land distribution. Altitude is an important aspect of physiography and has bearing on climate. Therefore, sample villages have been chosen on the basis of altitudinal and socio-ecological variations. Thus, five valleys namely upper Satlej, Tirung, Hangarang, Baspa and Ropa were selected. It was attempted to take one village from each valley falling under each altitudinal zone. However, entire Ropa and Hangarang valleys are located quite high, hence, there was no village situated at lower zone. Villages having less than twenty households were not taken up because such small villages blow up averages. The selected villages fall under the following altitudinal zones:

- The Lower Zone (Below 2200 mts)
- The Middle Zone (2200-2900 mts)
- The Higher Zone (2900-3900 mts).

Primary data was generated from 464 households from thirteen selected villages. There is lot of variations in the size of villages; extremely small villages, which have less than twenty households, were left out. At least fifty percent of households were surveyed in small villages that have less than hundred households whereas in bigger villages with hundred or more households, minimum of thirty percent households were surveyed. Among above households surveyed in sample villages, it was attempted to include fifty percent of poor households in each case.

Conventionally, all households in Kinnaur owned agricultural land. Only migrant workers are agricultural labourers. Agricultural activities are not only difficult but also yield less. Winter is agriculturally the off-season, hence, it requires people to store food for winter months. The system of community labour facilitates people to cultivate more than what a family is usually capable doing. This has brought more land under plough. The pressure due to ever-increasing population seeks to reclaim more land for agriculture. Terraced agricultural fields of different sizes and shapes are located mostly in lower and mid altitude. These areas have gentler gradient and have deeper soil cover. Similarly, agricultural fields are fewer and highly segmented on the northern slope due to a steeper gradient that results in thinner soil cover and lesser sunlight.

All 464 surveyed households own agricultural land. However, apart from cultivation, some households set aside some land, which is left to let grass grow. These two put together form total land holding. However, the emphasis here is on agriculture, so it includes only operational land holdings.

Tables 3 and 4 show that about 39 percent of surveyed households have got some portion of the cultivated land acquired since 1980. Out of these, forty-nine percent have already been officially transferred in the name of heads of respective households; whereas sanctions of ownership of about fifty percent cases are under process. The procedure of sanctioning new piece of land for agricultural operations is cumbersome and time consuming. The forest department keeps records of all land except private properties, which are registered with the revenue department. Land sought for cultivation needs to be cleared by the forest department and transferred to the department of revenue before it can be registered in the name of a private party.

Table 1
Distribution of occupied (purchased) land

Alt. zone	*Within villages*	*outskirts*	*Pastures*	*Total*
Higher zone	0	5	9	14
% of occupied land within H. Zone	0.0	35.7	64.3	100.0
% within each category columns	0.0	12.8	20.0	15.7
Middle zone	3	7	19	29
% of occupied land within M. Zone	10.3	24.1	65.5	100.0
% within each category columns	60.0	17.9	42.2	32.6
Lower zone	2	27	17	46
% of occupied land within L. Zone	4.3	58.7	37.0	100.0
% of occupied land within All Zone	40.0	69.2	37.8	51.7
All zone	5	39	45	89
% of occupied land within All. Zone	5.6	43.8	50.6	100.0
% within each category columns	100.0	100.0	100.0	100.0

Source: Field Survey.

The researcher was informed that these kinds of land transfer deals usually take about 5 to 6 years. However, people start agricultural operations in the meantime. Thus, there is unrecorded cultivated land too. Out of the all such land transferred, about seventy two percent ownership of land was awarded to the residents of lower and middle zones. Among all newly sanctioned operational agricultural fields, about fifty percent (Table 1) are on pastures and forty three percent on village outskirts. Only about five percent such sanction of land took place within villages. These trends clearly show that there is greater population pressure on land in lower and middle altitudinal areas than in the higher parts. But there is lesser land to meet people's demand for agricultural land in villages. People are moving out to village outskirts for their agricultural operations. Even this is not sufficient, and hence, even pastoral land has been acquired. As there is more pressure on land in Middle and lower zones, it not surprising that of all land purchased at pastures, about fifty percent belongs to villagers of the lower zone. Thus, a lot of marginal land has been transferred for regular agricultural purposes.

Table 2
Distribution of occupied (claimed) land

Alt. zone	*Within villages*	*outskirts*	*Pastures*	*Total*
Higher zone	1	3	21	25
% of occupied land within H. Zone	4.0	12.0	84.0	100.0
% within each category columns	100.0	50.0	24.4	26.9
Middle zone	0	1	36	37
% of occupied land within M. Zone	0.0	2.7	97.3	100.0
% within each category columns	0.0	16.7	41.9	39.8
Lower zone	0	2	29	31
% of occupied land within L. Zone	0.0	6.5	93.5	100.0
% of occupied land within All Zone	0.0	33.3	33.7	33.3
All zone	1	6	86	93
% of occupied land within All. Zone	1.1	6.5	92.5	100.0
% within each category columns	100.0	100.0	100.0	100.0

Source: Field Survey

Out of households cultivating land that has not been officially sanctioned since the 1980s, about ninety two percent (Table 2) cultivate pastureland. It shows that there is scarcity of land not only within villages but also on village outskirts. About seventy five percent of all the landholdings claimed at pastures across all three altitudinal zones are by households of lower and middle altitude. With increasing population pressure and lower scope of expansion of agricultural land in and around villages of lower and middle altitude, people are increasingly heading for nearest pastures and other marginal land for agricultural activities.

Plate 1: **Labourers working at a newly carving out agricultural farmland out of forest area near Chakka pasture at Kinnaur**

Nearly forty percent of all land claimed through official sanctions are by households of middle zone. This proportion is slightly lower for lower zone at 33 percent. This is probably because pressure on land and subsequent demand for agricultural holdings rose up much earlier in lower zone. So, the people of lower zone initiated land sanction processes earlier and got comparatively better land transferred, leaving only the limited possibility of acquisition of such lands for others. Table 3 shows that greater number of land sanctions took place for the households of lower zone. Contrary to this, there is relatively greater availability of land in middle zones. The same holds good for higher zones, which account for nearly 27 percent of land sought for ownership sanctions. Availability of newer agricultural fields for households is greater in higher zones due to low population. Fifty percent of sanctioned land is confined to village outskirts and only about 24 percent in pastures. It suggests that there is lower population pressure in higher zone.

Table 3

Size-wise distribution of Land Purchased since 1980s

Land Purchased	*Upto 1 Bigha*	*1-3 Bigha*	*3-5 Bigha*	*Above 5 Bigha*	*Total*
Higher zone	0	2	7	5	14
% all occupied land	0.0	14.3	50.0	35.7	100.0
% in the same holding size	0.0	11.1	14.3	23.8	15.7
Middle zone	1	8	15	5	29
% all occupied land	3.4	27.6	51.7	17.2	100.0
% in the same holding size	100.0	44.4	30.6	23.8	32.6
Lower zone	0	8	27	11	46
% all occupied land	0.0	17.4	58.7	23.9	100.0
% in the same holding size	0.0	44.4	55.1	52.4	51.7
All zone	1	18	49	21	89
% all occupied land	1.1	20.2	55.1	23.6	100.0
% in the same holding size	100.0	100.0	100.0	100.0	100.0

Source: Field survey

Plate 2: **A Ploughed agricultural field at forest land above Kalpa village**

Table 4

Size-wise distribution of land occupied since 1980s

Reclaimed Land	*Upto 1 Bigha*	*1-3 Bigha*	*3-5 Bigha*	*Above 5 Bigha*	*Total*
Higher zone	0	6	17	2	25
% all occupied land	0.0	24.0	68.0	8.0	100.0
% in the same holding size	0.0	16.7	36.2	50.0	26.9
Middle zone	3	16	17	1	37
% all occupied land	8.1	43.2	45.9	2.7	100.0
% in the same holding size	50.0	44.4	36.2	25.0	39.8
Lower zone	3	14	13	1	31
% all occupied land	9.7	45.2	41.9	3.2	100.0
% in the same holding size	50.0	38.9	27.7	25.0	33.3
All zone	6	36	47	4	93
% all occupied land	6.5	38.7	50.5	4.3	100.0
% in the same holding size	100.0	100.0	100.0	100.0	100.0

Source: Field survey

It is evident from the Table 4 that none of the households in higher zone have acquired less than one bigha of agricultural land. Among newly acquired agricultural land, about 21 percent measure up to 3 bighas across all zones while fifty percent is between 3 to 5 bighas. About 78 percent of all households that have purchased agricultural land have acquired more than three bigha. About 44 percent of newly purchased land measures up to 3 bigha of all acquired land in middle and lower zones. The figure for the same is only eleven percent in upper zone. This shows more unoccupied land available in higher zone compared to middle and lower zones. Highest proportion of land tilled without award of ownership sanctions measures 3 to 5 bighas in middle and higher zones (Table 4) whereas 38 percent is of 1 to 3 bighas in lower zone.

About 56 percent of surveyed households (Table 5) own agricultural land at pastures. These pieces of land are either officially sanctioned or are appropriated by households even without sanctions. Some households cultivate land at pastures both with and without sanctions. About 32 percent of households have been occupying pastureland for the past three decades; nearly 56 percent of such land belongs to households of lower zone. Lower zone together with middle zone account for 81 percent of pastureland owned or occupied prior to 1980.

Table 5
Distribution of land holding in pasture

	Upto 1 Bigha	*1-3 Bigha*	*3-5 Bigha*	*Above 5 Bigha*	*Total*
1	*2*	*3*	*4*	*5*	*6*
Occupied prior to 1980					
Higher zone	3	9	2	14	28
% all occupied land	10.7	32.1	7.1	50.0	100.0
% in the same holding size	11.5	30.0	5.1	26.9	19.0

(Contd...)

1	*2*	*3*	*4*	*5*	*6*
Middle zone	11	3	5	17	36
% all occupied land	30.6	8.3	13.9	47.2	100.0
% in the same holding size	42.3	10.0	12.8	32.7	24.5
Lower zone	12	18	32	21	83
% all occupied land	14.5	21.7	38.6	25.3	100.0
% in the same holding size	46.2	60.0	82.1	40.4	56.5
All zone	26	30	39	52	147
% all occupied land	17.7	20.4	26.5	35.4	100.0
% in the same holding size	100.0	100.0	100.0	100.0	100.0
Occupied Since 1980					
Higher zone	0	4	17	7	28
% all occupied land	0.0	14.3	60.7	25.0	100.0
% in the same holding size	0.0	12.5	28.8	46.7	23.9
Middle zone	8	15	23	3	49
% all occupied land	16.3	30.6	46.9	6.1	100.0
% in the same holding size	72.7	46.9	39.0	20.0	41.9
Lower zone	3	13	19	5	40
% all occupied land	7.5	32.5	47.5	12.5	100.0
% in the same holding size	27.3	40.6	32.2	33.3	34.2
All zone	11	32	59	15	117
% all occupied land	9.4	27.4	50.4	12.8	100.0
% in the same holding size	100.0	100.0	100.0	100.0	100.0
Occupied Till 2007-08					
Higher zone	3	13	19	21	56
% all occupied land	5.4	23.2	33.9	37.5	100.0
% in the same holding size	8.1	21.0	19.4	31.3	21.2
Middle zone	19	18	28	20	85
% all occupied land	22.4	21.2	32.9	23.5	100.0
% in the same holding size	51.4	29.0	28.6	29.9	32.2
Lower zone	15	31	51	26	123
% all occupied land	12.2	25.2	41.5	21.1	100.0
% in the same holding size	40.5	50.0	52.0	38.8	46.6
All zone	37	62	98	67	264
% all occupied land	14.0	23.5	37.1	25.4	100.0
% in the same holding size	100.0	100.0	100.0	100.0	100.0

Source: Field survey

It can be inferred from the table that demands for agricultural land rose sharply in lower zone due to greater population pressure. This was followed by middle zone. Fifty percent of households in higher zone occupied more than five bighas of pastureland prior to 1980. This figure goes down to 47 percent and 25 percent in middle zone and lower zone respectively. As discussed earlier, this suggests decreasing population pressure and greater availability of land with increasing height.

Proportion of landholdings occupied since the 1980s are much higher in middle and lower zone in comparison to upper zone. Forty two percent households of middle zone cultivate in pastures while the share is 34 percent in lower zone. As discussed earlier, it is probably because lower zone started experiencing population pressure much before middle zone and many households in lower zone had already occupied pastureland. On the contrary, middle zone started experiencing greater population pressure after 1980s, and had more pastureland available to be put to cultivation.

Land occupied till 2007-08 is about 46 percent in lower zone and 32 percent in middle zone. These two zones together comprise about 78 percent of all agricultural land occupied in pastures. About 62 percent of total land holding at pastures was of more than three bighas in size prior to 1980s. Even lower zone had about 62 percent holdings of 3 bighas and above out of all land holdings at pastures. Due to increasing demand for land, much of the land was already occupied prior to 1980s; there were fewer suitable land left at pastures. People turned to other marginal lands. Hence, the figure for lower zone declined to 32.5 percent. In comparison to lower zone, middle zone experienced population pressure later and it does not seem to be there in higher zone. Hence, there is increase in pastureland occupation in these two zones prior to 1980s and even after 1980s.

Table 6
Location of Land Applied by the households in about a decade

Altitudinal Zones	*Within Village*	*Outside Village*	*Pastures*	*Total*
Higher zone	1	3	10	14
% all applied land	7.1	21.4	71.4	100.0
% in the same holding size	16.7	8.3	25.0	17.1
Middle zone	3	4	14	21
% all applied land	14.3	19.0	66.7	100.0
% in the same holding size	50.0	11.1	35.0	25.6
Lower zone	2	29	16	47
% all applied land	4.3	61.7	34.0	100.0
% in the same holding size	33.3	80.6	40.0	57.3
All Zone	6	36	40	82
% all applied land	7.3	43.9	48.8	100.0
% in the same holding size	100.0	100.0	100.0	100.0

Note: Row % is percentage of each figure to the total of all figures in the same row which is 100 per cent. Cloumn% is percentage of each figure to the total of all figures in the same column which is 100 per cent.

Source: Field survey.

It can be seen from the Table 6 that the pressure on land is increasing along with the demand for agricultural land. Nearly seventeen percent of the surveyed households have applied for a grant of new agricultural land in the last 5 to 10 years. Households of lower zone constitute about 57 percent of such applicants while 25.6 percent and 17.1 percent applicants are from middle and higher zone respectively. This shows greater population pressure, hence, greater demand for land as we go lower. Nearly 49 percent of the land cited by applicants is located in pastures, about

44 percent in village outskirts and 7 percent within villages. Thus, it can be said that more marginal land, which is steeper and vulnerable, will be brought under plough in future. A greater proportion of households in higher and middle zones want land in pasture while a larger number of applicants in lower zone have cited land from village outskirts. This is because in lower zone there is little agricultural land available in pastures. Even going too high on pastureland is not an option because this would seriously restrict the growing season. Therefore, even the difficult slopes of village outskirts along valleys etc. are sought after in the lower zone.

The above analysis suggests that the expansion of agricultural land started earlier by households of the lower zone. Population pressure occurred there earlier than middle and higher zones. Households of middle zone and higher zone followed this when their population and needs started increasing. The expansion of agricultural land at the expense of marginal land reduced grass-growing area. As there is higher population pressure in lower zones than middle and higher zones, there is greater demand of agricultural land in this zone. But there is little availability of land for agriculture in and around villages. Even pasturelands are at stress. That is why an increasing number of hazard prone slopes in village outskirts are being transformed into terraced agricultural fields in lower zone. Such lands are unable to produce good agricultural harvests after a couple of years. This in turn gives rise to more demand for land, even on steep and vulnerable slopes.

Distribution of Agricultural Land

The total area available for agriculture is an important factor in any mountainous region of harsh environment. Relying solely on technological inputs to meet the demands thrown up by increased population pressure is not only difficult but also very expensive. In remote mountainous areas, where technological resources are generally low level, land holding size acquires great importance. The size of land holding is roughly related to production. Keeping this in mind, the surveyed households have been divided into following five groups as per size of operational land:

i) Up to three bigha of land = Marginal Farm/Farmer

ii) Three to five bigha of land = Small Farm/Farmer

iii) Five to ten bigha of land = Medium Farm/Farmer

iv) Ten to fifteen bigha of land = Large Farm/Farmer

v) Above fifteen bigha = Very Large Farm/Farmer.

In order to analyse the relationship between the pressure on the different sized land holdings and the persons dependent on the respective land holdings, households were grouped according to the number of family members. The following four groups of family members were thus formulated:

i) Up to 4 members = Small Family

ii) 5 to 7 members = Medium Family

iii) 8 to 10 members = Large Family

iv) Above 10 = Very Large Family.

Though there are variations between lower and middle zones in the proportion of farmers, but the trends are almost the same as that of the proportions for overall figures. Higher population pressure in lower zone has resulted in greater fragmentation of land. Besides, the preference for nuclear family in recent decades has also contributed to the fragmentation of land. About 38 percent famers in higher zone are large and very large while the same figure for middle zone and lower zone are 18 percent and 11.2 percent respectively. The reason behind this difference is the lower incidence of fragmentation of land as we go up. Higher areas have stronger hold of tradition [this means as we climb higher, the tradition and culture becomes stronger], hence, more number of joint families and lesser division of farm states. In general, lower areas are more populous than the higher areas.

As discussed previously, more suitable parts of land have already been put under plough and now even steeper slopes are worked upon in lower zones. The fact that farmers are carving out agricultural terraces on hazardous slopes and the greater proportions of small and medium farmers suggest that land fit for agriculture is getting scarcer in lower and middle zones. A better assessment of pressure on agricultural land can be analyzed in terms of family size and operational holdings. Smaller families with larger land holdings have lower pressure on agricultural land while lager families with smaller operational holdings have higher pressure.

It is clear from table 7 that the highest proportion of small families amounting 33.9 percent have small farms while 31.2 percent of small families are of medium farmers, followed by 26.5 percent having less than three bigha of land and are marginal farmers. Among medium families, highest proportion of 36.6 percent has land between 5 to 10 bigha. About 31.7 percent families of this class have land between 3 to 5 bigha; figures for marginal, large and very large farmers in the same family class are 17.4, 10.6 and 3.7 percent respectively. Similarly, 34.2 percent of large families have medium sized operational holdings followed by large farmers who constitute about 31 percent of all large families. In case of very large families, 41.5 percent have medium farms followed by 36.6 and 22 percent with large and very large farms respectively.

It can be deduced from above discussion that there is no direct linkage between family size and farm size. Small and medium families constitute biggest proportions that have small and medium sized operational land holdings. On the other hand, large and very large families mostly have medium to large sized agricultural holdings. It can be concluded that a considerable proportion of small families with smaller holdings and larger families with medium farms are facing problems.

Table 7
Distribution of Operational Landholdings

	Marg. Farm	Small Farm	Medium Farm	Large farm	V. Large Farm	Total
Family size in Higher Zone						
Small Family	1	13	21	7	1	43
% of all farm size	2.3	30.2	48.8	16.3	2.3	100.0
% within same farm size	50.0	38.2	61.8	22.6	8.3	38.1
Medium Family	1	17	9	5	4	36
% of all farm size	2.8	47.2	25.0	13.9	11.1	100.0
% within same farm size	50.0	50.0	26.5	16.1	33.3	31.9
Large Family	0	4	3	11	5	23
% of all farm size	0.0	17.4	13.0	47.8	21.7	100.0
% within same farm size	0.0	11.8	8.8	35.5	41.7	20.4
Very Large Family	0	0	1	8	2	11
% of all farm size	0.0	0.0	9.1	72.7	18.2	100.0
% within same farm size	0.0	0.0	2.9	25.8	16.7	9.7
Total	2	34	34	31	12	113
% of all farm size	1.8	30.1	30.1	27.4	10.6	100.0
% within same farm size	100.0	100.0	100.0	100.0	100.0	100.0
Family size in Lower Zone						
Small Family	37	31	22	6	1	97
% of all farm size	38.1	32.0	22.7	6.2	1.0	100.0
% within same farm size	67.3	55.4	33.8	31.6	25.0	48.7
Medium Family	17	22	25	3	1	68
% of all farm size	25.0	32.4	36.8	4.4	1.5	100.0
% within same farm size	30.9	39.3	38.5	15.8	25.0	34.2

(Contd...)

	Marg.	*Small Farm*	*Medium Farm*	*Large Farm*	*V. farm*	*Total Large Farm*
Large Family	1	3	11	5	1	21
% of all farm size	4.8	14.3	52.4	23.8	4.8	100.0
% within same farm size	1.8	5.4	16.9	26.3	25.0	10.6
Very Large Family	0	0	7	5	1	13
% of all farm size	0.0	0.0	53.8	38.5	7.7	100.0
% within same farm size	0.0	0.0	10.8	26.3	25.0	6.5
Total	**55**	**56**	**65**	**19**	**4**	**199**
% of all farm size	27.6	28.1	32.7	9.5	2.0	100.0
% within same farm size	100.0	100.0	100.0	100.0	100.0	100.0
Family size All Zone						
Small Family	50	64	59	14	2	189
% of all farm size	26.5	33.9	31.2	7.4	1.1	100.0
% within same farm size	62.5	49.2	36.9	20.3	8.0	40.7
Medium Family	28	51	59	17	6	161
% of all farm size	17.4	31.7	36.6	10.6	3.7	100.0
% within same farm size	35.0	39.2	36.9	24.6	24.0	34.7
Large Family	2	15	25	23	8	73
% of all farm size	2.7	20.5	34.2	31.5	11.0	100.0
% within same farm size	2.5	11.5	15.6	33.3	32.0	15.7
Very Large Family	0	0	17	15	9	41
% of all farm size	0.0	0.0	41.5	36.6	22.0	100.0
% within same farm size	0.0	0.0	10.6	21.7	36.0	8.8
All Zone	80	130	160	69	25	464
% of all farm size	17.2	28.0	34.5	14.9	5.4	100.0
% within same farm size	100.0	100.0	100.0	100.0	100.0	100.0

Source: Field survey.

In higher zone, none of the very large and large families have marginal farm while maximum proportions of small families have medium to small farms. A majority of large families and very large families possess large or very large operational holdings.

The highest proportion of small families in the middle zone has small sized land holdings and none of the small families own very large operational landholding. About 44 percent of medium sized families have medium sized farm. Even large and very large sized families mostly have medium sized farms. Most very large families, have medium sized farms, followed by very large sized farms. In this zone, none of the very large families have marginal or small farms. About 69 percent of large families consist of medium, small and marginal farmers. Small families own significant land, however, population pressure on land holdings of medium and large families are higher in middle zone.

Lower zone have highly fragmented landholdings; 93 percent smaller families in lower zone consists of marginal, small and medium farmers. Among these, marginal farmers with 38.1 percent constitute the largest group followed by small farmers who account for 32 percent. In case of medium families, marginal and small farmers make up to 57.4 percent. However, medium farmers with 36.8 percent constitute the largest group among medium families. The large family group has 52.4 percent medium farmers followed by 23.8 percent of large farmers who own ten to fifteen bigha of land. None of the very large families among surveyed households are marginal or small farmers. In fact, 93.3 percent of very large families are medium to large farmers; here very large families constitute 58.9 percent of medium farmers alone.

Conclusion

The overall discussion shows that smaller families are exerting less pressure on agricultural resources. Contrary to this, larger families are putting agricultural resources at greater stress. Since there is more population in middle and lower zones, these experience [exert] higher pressure on land. In other words it can be said that with the maximum population, lower zone has highest pressure and higher zone has lowest pressure due to least population among the three zones.

Land is a limited resource; it is more so in a mountainous region like Kinnaur, which has extremely rugged topography and harsh climatic conditions. Therefore, there is a limited possibility of putting non-agricultural land under plough. In order to meet growing food requirements, cultivating two crops in a year wherever it is feasible is necessary. Mountain soils in general are poor, so for raising two crops, certain technological inputs such as high yield varieties (HYV) seeds, chemical fertilizers, pesticides, insecticides and better irrigation facilities are needed.

References

Rawat, A.S. (2002). *Alternative Farming Systems in the Dry Temperate Zone of Himachal Pradesh: A Case Study of Kinnaur District*. New Delhi: Indus Publishing Company.

Sarkar, Rinki (1998). Understanding Sustainability: Study of a Hill Settlement, *Economic and Political Weekly,* Vol. 33, No. 44 (Oct. 31 - Nov. 6, 1998), pp. 2800-2806.

Purohit, A.N. (1994). In the Foreword, in Y.P.S. Pangtey and R.S. Rawal (eds), *High Altitudes of the Himalaya: Biogeography, Ecology and Conservation*. Nainital: Gyanodaya Prakashan.

Sharma, P.D. and S. Minhas (1993). 'Land use and bio physical environment of Kinnaur district. Himachal Pradesh, India', *Mountain Research and Development*, 13(1): pp. 41–60.

Chandra, Ramesh (1992). *Highlanders of North Western Himalayas*. (Tribal studies of India Series T153). New Delhi: Inter-India Publications.

Singh, Harjit (1992). 'Ecological Set-up and Agrarian Structure in High Altitude Villages of Ladakh', in R.B. Singh (ed.). *Dynamics of Mountain Geo-Systems*, pp. 204–221. New Delhi: Ashish Publishing House.

Gupta, Ramji (1992). 'Ecological Setting of the Central and Western Himalayas' in M.K. Raha and P.C. Coomar (eds). *The Himalayas and the Himalayans (Studies on Ecology, Biology and Culture)*, pp. 3-10. Calcutta: Anthropological Survey of India.

Karan, P. Pradyumna (1989). Environment and Development in Sikkim Himalaya: A review, *Human Ecology,* Vol. 17, No. 2, Human Ecology in the Himalaya , pp. 257-271, Springer

Chib, S.S. (1987). *Kinnauras of Trans Himalayas*. New Delhi: Light and Life Publishers.

Raha, M.K. and S.N. Mahto (1985). *Kinnaurese of the Himalayas*. Calcutta: Anthropological Survey of India.

Schroeder, R.F (1985). Himalayan Subsistence system: Indigenous agriculture in Rural Nepal, *Mountain Research and Development,* Vol.5, No. 1 pp. 31-44.

Bajpai, S.C. (1981). *Kinnaur in the Himalayas: Mythology to Modernity*. New Delhi: Concept Publishers.

Mather, Cotton & P. P. Karan (1978).*The Concept of Land Support Units: Bhutan*, Geografiska Annaler. Series B, *Human Geography,* Vol. 60, No. 1 (1978), pp. 28-35, Swedish Society for Anthropology and Geography, Blackwell Publishing.

6

Medicinal Plant Biodiversity in the Darjeeling Himalaya

Dhani Raj Chhetri

Introduction

Darjeeling Himalaya is situated between the 87° 59′-88° 53′ E and 28° 31′-27° 13′N in the Eastern Himalayas of India. It is the frontier district running up between Nepal and Bhutan, and stretching from the plains of Bengal in the south to Sikkim in the North. It is bordered by Bhutan in the East and Nepal in the West. The Three Hill sub-divisions of Darjeeling District, Kalimpong, Kurseong and Darjeeling Sadar, consisting of eight developmental blocks, and occupying an area of 2417 km^2 comprise the Darjeeling Himalaya (Fig-1). Due to their great variation, a wide array of climatic zones is available (Chhetri et al., 2005a, b) that favor the growth of its peculiar flora and fauna.

The region of Darjeeling Hills displays several diverse factors, giving rise to the following three major climatic zones:

a. Sub-tropical region or lower hill zone
b. Sub-temperate region or middle hill zone
c. Temperate or upper hill zone.

Fig 1.: Darjeeling Himalayan Region

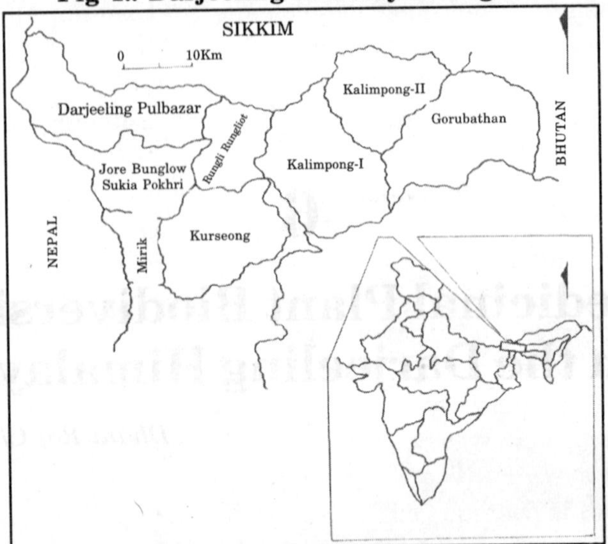

Within these three geographical regions there is a wide range of variation in altitude, temperature, humidity, photoperiod and soil conditions, which have contributed to the existence of a wide variety of flora. Floristic studies suggest a good deal of genetic diversity and endemism among the medicinal and aromatic plants in North-east Himalayas. In terms of medicinal plant biodiversity, Darjeeling Himalayas, situated in one of the biodiversity hot-spots of the world (Eastern Himalaya) is a source of untold riches which we cannot neglect.

More than 90% of medicinal plants in the hills are found in the natural forests. The Himalayan forest in Darjeeling hills is distributed at the altitudinal range from 200-3660 m. The soil in Darjeeling Himalaya is mostly brownish, sandy loam, which is slightly acidic and rich in organic matter. Most of the hill slopes receive full blast of southwest monsoon with annual rainfall ranging from 2000-6000 mm. Vegetation varies with altitude, aspect and soil type. The forest can be generally divided as follows:

Lower hill forest (upto-1000m)

Contains *Shorea robusta and Tectona grandis* in pure patches mixed with other important species like *Terminalia myriocarpa, Schima wallichii, Cedrella toona,Michelia spp.* etc.

Middle hill forests (1000-2000m)

Represented by *Betula spp., Quercus spp., Juglans regia, Castanopsis spp.* etc.

Upper hill forests (above-2000m)

Represented by *Alnus nepalensis, Acer spp., Machilus spp., Magnolia spp.* etc. In the alpine region Rhododendron and conifers like *Tsuga dumosa, Abeis webbiana, Taxus wallichiana* are the dominant vegetation.

Medicinal Plant Biodiversity and Traditional Medicine

India is the world's 12th mega biodiversity center, having over 45,000 plant species of which about 18,000 are the flowering plants. Though the Himalayas cover only 18% of India's geographical area, it accounts for more than 50% of the forest cover and 40% of the species endemic to the Indian subcontinent. Depletion in the forest cover and ecosystem services are interlinked and threaten the sustainable livelihood of not only hill people but also the much larger population inhabiting the adjoining Indo-Gangetic plains (Hamilton, 1987; Hurni, 1999).

The Indian Himalayan region itself is a rich repository of medicinal plants with a total of 1784 species. People living in interior villages depend completely on forest resources for maintaining their day-to-day needs like food, fuel, household articles and medicines. With only 8 hospitals and 24 primary health centres, modern medical facilities are rare in the Darjeeling hills. The doctor to people ratio is 1:4892 (Chhetri, 2004). Therefore, people are largely dependent on herbal medicine for their healthcare. The lack of hospitals in this region forces people to follow these ethnic traditions.

Countries like Africa, Asia and Latin America use traditional medicine to help meet some of their healthcare needs. In Africa up to 80% of the population uses traditional medicine for primary healthcare. In industrialized countries adaptations of traditional medicine are termed complementary and alternative. At least 25% of modern medicine is of plant origin. A huge chunk of humanity is still dependent on herbal medicine for their healthcare needs. In India, traditional health care systems like Ayurveda, Unani and Siddha operate alongside modern healthcare. There are 400, 000 registered traditional medicine practitioners in India compared to 332,000 registered doctors. Darjeeling hills in one of remote regions of the country and being placed in one of the biodiversity hot-spots of the world is a virtual treasure trove of medicinal plant biodiversity (Table-1).

Table 1
The most commonly used medicinal plants found in Darjeeling Himalayas

Botanical Name of the plants	*Botanical Name of the plants*
Abeis webbiana	*Bergenia sp.*
Abroma augusta	*Betula utilis*
Abrus precatorius	*Bidens pilosa*
Abutilon indicum	*Bischofia javanica*
Achillea mellifolium	*Boehmeria rugulosa*
Achyranthus aspera	*Botrychium ternatum*
Achyranthus aspera	*Bridillia retusa*
Aconitum sp.	*Buddleja asiatica*
Acorus calamus	*Calamintha umbrosa*
Actinodaphne hookeri	*Callicarpa arborea*
Adiantum capillus-veneris	*Campylandra aurantica*
Adina cordifolia	*Cassia fistula*
Ageratum conyzoides	*Castanopsis sp.*
Agrimonia sp.	*Cautleya sp.*
Ajuga bracteosa	*Celastrus paniculata*

(Contd...)

Botanical Name of the plants	*Botanical Name of the plants*
Allium sp	*Celosia argentea*
Alnus nepalensis	*Cephalis ipecacuanha*
Aloe barbadensis	*Cephalotaxus sp.*
Alpinia galangal	*Cestrum sp.*
Alstonia scholaris	*Choerospondias axillaris*
Androsace sarmentosa	*Chrysanthemum sp.*
Anemone sp.	*Cinnamomum tamala*
Anthogonium gracile	*Cirsium bolocephalum*
Aphanamixis polystachya	*Cissampelos pareira*
Aquilegia vulgaris	*Citrulus colocynthis*
Arisaema sp.	*Citrus sp.*
Aristolochia indica	*Clematis sp.*
Artemisia sp.	*Codonopsis ovata*
Arundinaria falcata	*Coelogyne sp.*
Asandra butyracea	*Coleus sp.*
Asparagus racemosus	*Commelina bengalensis*
Asplenium falcatum	*Corydylis govaniana*
Aster sp.	*Costus speciosus*
Astilbe rivularis	*Cryptocarya amygdalina*
Astragalus sp.	*Curculigo orchoides*
Bambusa arundinaria	*Curcuma sp.*
Bauhinia sp.	*Cuscuta sp.*
Begonia picta	*Cyanoglossum zeylanicum*
Berberis sp.	*Cymbidium aloefolium*
Cyathuala prostrata	*Hemiphragma heterophyllum*
Daphne cannabina	*Heracleum sp.*
Datura sp.	*Hibiscus rosa-sinensis*
Delphinium sp.	*Hippophae sp.*
Dendrobium macraei	*Holarrhena antidyssentrica*
Dendrocalamus sp.	*Holboellia latifolia*
Dendropthoe falcata	*Houttuynia cordata*
Dicentra thalictrifolia	*Hoya lanceolata*
Dichroa febrifuga	*Hydrocotyl sp.*
Digitalis sp.	*Hymenodictyon excelsum*
Dioscorea sp.	*Hyocyamus niger*
Dolichos biflorus	*Hypericum sp.*
Drymaria cordata	*Impatiens balsamina*
Dryopteris filix-mas	*Imperata cylindrical*
Eclipta prostrata	*Ipomoea sp.*
Edgworthia gardneri	*Iris sp.*

(Contd...)

Botanical Name of the plants	*Botanical Name of the plants*
Elaeocarpos sp.	*Jatropa curcas*
Elastostemma lineolatum	*Juglans regia*
Elsholtzia blanda	*Juniperus sp.*
Entada scandens	*Justicia gendarus*
Ephedra sp.	*Kaempferia rotunda*
Epilobium latifolium	*Laportea terminalis*
Equisetum sp.	*Leea robusta*
Erigeron sp.	*Leucosceptrum canum*
Erythrina arborescens	*Ligularia amplexicaulis*
Eulophia campestris	*Lindera neesiana*
Eupatorium cannabianum	*Litsea sp.*
Euphorbia sp.	*Loranthus falcata*
Euphrasia officinalis	*Lycopodium sp.*
Eurya nitida	*Lyonia ovalifolia*
Evodia fraxinifolia	*Macaranga indica*
Fagopyrum sp.	*Machilus gammiena*
Ficus sp.	*Maesa chisia*
Fragaria vesca	*Maharanga emodi*
Fraxinus floribunda	*Mahonia acanthifolia*
Fritillaria sp.	*Mallotus philippinensis*
Fumaria perviflora	*Marsdenia roylei*
Garanga pinnata	*Meconopsis sp.*
Garcinia cowa	*Melastoma malabaricum*
Gaultheria fragrantissima	*Melissa axillaries*
Gentiana sp.	*Mesua ferrea*
Geranium nepalense	*Michelia sp.*
Gingko biloba	*Mimosa pudica*
Girardinia palmata	*Monstera deliciosa*
Gloriosa sp.	*Morina longifolia*
Glycosmis arborea	*Morus indica*
Guizotia abyssinica	*Mucuna monosperma*
Gynocardia odorata	*Mussenda sp.*
Hedera sp.	*Myrica esculenta*
Hedyotis scandens	*Myricaria rosea*
Helicia nilagirica	*Myristica fragrans*
Nardostachys jatamansi	*Primula sp.*
Nasturtium fontana	*Prunella vulgaris*
Nepeta discolor	*Prunus cerasoides*
Nephrolepia cordifolia	*Pterigota alata*

(Contd...)

Botanical Name of the plants	*Botanical Name of the plants*
Nicrandra physaloides	*Pteris sp.*
Ophioglossum vulgatum	*Pterocarpos marsupiam*
Orchis latifolia	*Quercus sp.*
Origanum vulgare	*Ranunculus sp.*
Origanum vulgare	*Raulvolffia serpentina*
Orobranche sp.	*Rhamnus sp.*
Orobranche sp.	*Rheum sp.*
Oroxylum indicum	*Rhodiola sp.*
Osbeckia sp.	*Rhododendron sp.*
Ostodes paniculata	*Rhus sp.*
Osyris arborea	*Roscoea capitata*
Oxalis corniculata	*Rubia sp.*
Paedaria foetida	*Rubus sp.*
Panax psedudo-ginseng	*Rumex nepalensis*
Pandanus nepalensis	*Salix babylonica*
Paris polyphylla	*Sambucus nigra*
Parthenocissus semicordata	*Sapindus mukurossi*
Passiflora nepalensis	*Sarcopyramis nepalensis*
Pavetta indica	*Saussurea sp.*
Pedicularis sp.	*Schima wallichii*
Pentapanax leschenaulti	*Selenum tenuifolium*
Peperomia sp.	*Semicarpos zeylanica*
Perilla frutiscens	*Senecio scandens*
Persia americana	*Siegesbeckia orientalis*
Persicaria chinense	*Silene sp.*
Phlogacanthus pubinervius	*Skimmia laureola*
Pholidota articulate	*Smilax zeylanica*
Phyllanthus emblica	*Solanum sp.*
Physales minima	*Sonchus arvensis*
Phytolacca acinosa	*Sphaeranthus indicus*
Picrorhiza kurrooa	*Spilanthes acmella*
Pieris ovalifolia	*Spondias mangifera*
Pilea microphylla	*Stellaria mediaS*
Pinus sp.	*tephania sp.*
Piper sp.	*Sterculia villosa*
Piricampylus glaucus	*Sterospermum tetragonatum*
Pittosporum sp.	*Streptolirion volubile*
Plantago sp.	*Swertia sp.*
Plumera rubra	*Sylvia officinalis*

(*Contd...*)

Botanical Name of the plants	Botanical Name of the plants
Plumeria acutifolia	*Symplocos racemosa*
Podophyllum hexandrum	*Tagetes patula*
Polygala arillata	*Taraxacum officinale*
Polygonatum sp.	*Taxus baccata*
Polygonum sp.	*Terminalia sp.*
Portulacca oleracea	*Thalaspi arvense*
Potentilla sp.	*Thalictrum foliolosum*
Premna sp.	*Thuja sp.*
Thysanolaena agrostis	*Viburnum mullaha*
Tinospora cordifolia	*Viola sp.*
Toddalia asiatica	*Viscum articulatum*
Trewia nudiflora	*Vitex negundo*
Tupistra nutans	*Vitis pedata*
Uncaria pilosa	*Walsura tribulata*
Urtica sp.	*Withania somnifrera*
Valeriana hardwickii	*Woodfordia fruticosa*
Verbascum thapsas	*Zanthoxylum sp.*
Vernonia anthelmentica	*Zingiber officinale*

The total medicinal plants of the Darjeeling Himalaya may be categorized into three different types.

1. *Cultivated plants of known pharmaceutical value*

The department of Cinchona and other medicinal plants has been cultivating this group of plants at three regions viz. Mungpoo, Munsong and Rongo-Gairibas. The plants are Cinchona, Ipecac, Foxglove, *Dioscorea sp. Rauwolfia serpentina* etc.

2. *Wild plants of known pharmaceutical value*

This category of medicinal plants have entered into cultivation to some extent but their number have been threatened in the wild due to rampant collection from the wild and exportation to pharmaceutical industries. This group is represented by *Aconitum bisma, A. spicatum, Gynocardia odorata, Litsea cubeba, Panax pseudoginseng, Phyllaanthus emblica, Rheum acuminatum, Rubia mantjith* etc.

3. Wild plants of known ethnomedicinal uses

A large number of plants in the Darjeeling Himalayas belong to this category. This group of medicinal plants has not so far been studied from modern scientific point of view but used extensively in different traditional medicine system of the locality. These plants are represented by *Artemisia dubia, Astilbe rivularis, Berginia ciliata, Dichroa febrifuga, Drymaria cordata, Heracleum wallichii, Juglans regia, Oroxylon indicum, Rhododendron arboretum, Swertia chirata* etc.

Translating Medicinal Plants into Wealth

Medicinal plants have a high potential for enhancing economic growth and creating additional jobs in resource constrained areas suffering from limited educational opportunities, lack of infrastructure and underdeveloped commercial activities. The conversion of socio-cultural traditions and indigenous knowledge into economic opportunities also has the advantage of preserving the vanishing cultural knowledge and practices, which are threatened due to globalization. As the Himalayas are recognized as a treasure trove of biological and cultural diversity, there is a need to protect indigenous knowledge and cultural diversity on an urgent basis. In the uplands of North-East Himalayas, the means of achieving this may be by providing economic value to the traditional and indigenous knowledge (Ramakrishnan, 1992).

It has been realized that medicinal plants of the Indian Himalayan region offer advantage in having much greater possibilities of providing new molecules used in drugs due to the extreme condition that they are subjected to. (Sharma et al., 1997). For the sake of drug development, instead of random search for medicinal plants, a selective search based on traditional knowledge would be more focused, productive and certainly more economically viable (Rameswararao et al., 2003).

About 25% of modern medicine is of plant origin, a number of which have been derived from Himalayan medicinal plants for example: Digitalin and Digoxin are heart drugs derived from foxglove, *Digitalis purpurea.* Taxol, a cancer drug is derived from *Taxus wallichiana* (Raghav, 2003).

A tremendous scope for research is foreseen on the biochemical, pharmacological and pharmaceutical aspects of folk medicine and its related plants in Darjeeling Himalayas. The rich medicinal plant biodiversity of the area and a rich traditional medicine system could be understood from the single fact that for a single disease there are multiple remedies in the Darjeeling Himalayas. For example, there are 31 species of plants used separately as antipyretic agents and 23 species used as hepatoprotective agents (Chhetri, 2005).

When considered with regard to food security, there is no scope to commit agricultural land for medicinal plants cultivation. However, an important case is made for the encouragement of village tribal, small and marginal farmers etc. to grow medicinal plants in their household garden, waste land, fallow land etc. instead of the big farmers. Such a strategy can invigorate the growth and development of medicinal plant sector by linking it to the economy of the rural and tribal people (Saini, 2002).

The Need to Conserve

Himalayan plants above the treeline are so diverse at the smallest scales, simply because of the small size of the species. One may find more than 30 plant species in an alpine fellfield on an area the size of this page. The reasons for high plant species diversity in mountains are many. On a global and regional scale, mountains are hot-spots of biological richness because of the rapid altitudinal change of climatic conditions over a very short distance (Korner, 2004). At a finer scale, mountains biodiversity mirrors the geodiversity resulting from topography. Exposure and inclination of slopes and relief lead to a multitude of microclimatic conditions, which in combination with the substrate types and the associated water and nutrient regimes, create a great variety of microhabitats each with its characteristic assemblage of organisms.

It is estimated that the flowering plants of Darjeeling District are represented by 4000 species belonging to 160 families. There are also 300 ferns of which 8 are tree fern species. Of the 140 species of liverworts recorded from Darjeeling 72% are endemic. 287 species of Orchids are found in Darjeeling of which 66 are horticulturally important. Despite all these, the biodiversity of Darjeeling is suffering on account of ill-planned developmental activities, defective agricultural practices, unregulated use of pesticides in the tea gardens, over-grazing and acute shortage of energy in Darjeeling. Conservation of biodiversity should be kept in mind while planning developmental activities, especially the ones that put pressure on forests.

40% of the Indian Himalaya is degraded land, which calls for reclamation. Rehabilitation will be a key intervention if it enables local communities to regenerate and conserve natural resources while at the same time aiding their social development. The ecological benefits gained from vegetation recovery in the Himalaya would also help the global community by reducing the frequency and intensity of flood damage, in the downstream region and improving biodiversity conservation and carbon sequestration.

Conservation of biodiversity, which nourishes the tribal culture and heritage, is also equally important, if traditional knowledge has to be preserved. Commercial collection of traditional medicinal plants from tribal habitats has not only threatened the biodiversity, but it has also become the cause of hastening the loss of traditional knowledge. Therefore, the important plants used in the traditional medicine should be preserved in the field gene-banks, the tribals should be encouraged to continue their herbal practices and fresh tribal youths should be trained in their ethnic herbal heritage.

Biopiracy has recently emerged as a major threat to this traditional knowledge. The episodes acquiring patent rights by US scientists on neem, turmeric and basmati rice are cases in point. With limited supply of natural resources in their own country, the developed nations turns to the third world countries like India for biological resources. The third world countries are generally not only endowed with rich biodiversity but are also equally rich in traditional knowledge and wisdom that can be used to generate economic and intellectual properties. The Himalayan region being rich in biodiversity as well as traditional knowledge will have to find ways and means to protect its traditional knowledge from biopiracy. Time and again, the economically advanced countries have generated patents on pirated traditional knowledge from different countries (Table-2).

Table 2

Some patent claims by advanced countries based on pirated traditional knowledge from third-world countries

Common name of the plant	*Botanical name*	*Country in which traditionally used*	*Patent claimed by*
Neem	*Azadirachta indica*	India	USA, Japan
Plao-noi	*Croton sublyratus*	Thailand	Japan
Haldi	*Curcuma longa*	India	USA
Hoodia	*Hoodia gordoni*	Africa	UK
Karela	*Momordica charantia*	India	USA
Basmati rice	*Oryza sativa*	India, Pakistan	USA
Kava	*Piper mythisticum*	Fiji, Vanuatu	France
Kothala-himbutu	*Salacia reticulata*	Sri Lanka	Japan
Snake gourd	*Trichosanthes kirilowii*	China	USA

In the backdrop of this scenario, it is felt that if proper research is not initiated we will be the loser in many counts like:

i). The traditional knowledge nurtured through centuries and passed on from generation to generation may be lost forever.

ii). Some of the world's most rare medicinal plants, available only in this region, may become extinct altogether.

iii). Some of the incurable diseases of mankind, the cure of which may have remained hidden in the unique traditional knowledge and the equally unique biodiversity of this region may not be discovered at all.

The Threatened Medicinal Plants

The medicinal plants used in the local health traditions are gradually becoming extinct or are threatened (Table-3) due to developmental activities and population explosion (Chhetri et al., 2005b). In order to reverse this trend, domestication of wild medicinal plants is of utmost importance. Appropriate research should be carried out in institutions in the hills to develop agro-techniques for the cultivation of medicinal plants on priority basis. In order to safeguard this knowledge, it should be documented, preserved and patented.

Climate Change and Mountain Biodiversity

A consistent increase in CO_2 concentration in the atmosphere is expected to have a strong influence on high altitude plant biodiversity. Surprisingly, the exposure of alpine grassland to almost twice the current CO_2 concentration did not stimulate biomass production, but different species responded differently with one species gaining in mass and abundance taht may translate into change in biodiversity in the long run (Korner et al., 1997). Further studies on CO_2 enrichment trial in the Swiss treeline showed no significant CO_2 effect on growth in dwarf shrubs and pine. However, the deciduous *Larix decidua* did show a consistent stimulation

Table 3
Threatened medicinal plants of the Darjeeling Himalaya

Botanical Name	*Family*	*Local Name*	*Life form*
Abutilon indicum	Malvaceae	Ghanti phool	Shrub
Aconitum spicatum	Ranunculaceae	Bikh	Herb
Aconitum heterophullum	Ranunculaceae	Bikhumma	Herb
Aconitum palmatum	Ranunculaceae	Bikhumma	Herb
Aeschynanthus sikkimensis	Gesneriaceae	Sinduray	Herb
Cinnamomum tamala	Lauraceae	Sinkauli	Tree
Cissampelos pareira	Menispermaceae	Batulay	Climber
Clematis buchaniana	Ranunculaceae	Pinasay lahara	Climber
Curcuma zeodaria	Zingiberaceae	Kalo hardi	Herb
Dactylorhiza hatagirea	Orchidaceae	Panch aunlay	Herb
Elaeocarpos granitus	Elaeocarpaceae	Rudraksha	Tree
Gloriosa superba.	Liliaceae	Bikhphool	Herb
Gynocardia odorata	Flacourteaceae	Gantay	Tree
Hedychium spicatum	Zingiberaceae	Dudh kewnra	Herb
Juglans regia	Juglandaceae	Okkhar	Tree
Mahonia acanthifolia	Berberidaceae	Chutro	Tree
Mesua ferrea	Clusiaceae	Nageswar	Tree
Nardostachys jatamansi	ValerianaceaE	Jatamaasi	Herb
Ophioglossum vulgatum	Ophioglossaceae	Jibray saag	Herb
Piper longum	Piperaceae	Pipla	Climber
Panax pseudoginseng	Araliaceae	Salanay	Herb
Picrorhiza kurroa	Scrophulariaceae	Kutki	Herb
Podophyllum hexandrum	Podophyllaceae	Papari	Herb
Rheum acuminatum	Polygonaceae	Padamchal	Herb
Rhododendron anthopogon	Ericaceae	Sunpatay	Shrub
Rhododendron arboreum	Ericaceae	Lali gurans	Tree
Rhododendron campanulatum	Ericaceae	Nilo chimal	Shrub
Rhus semialata	Anacardiaceae	Bhakimlo	Tree
Saussurea costus	Asteraceae	Kuth	Herb
Stephania glabra	Menispermaceae	Nemilahara	Climber
Stephania hernandifolia	Menispermaceae	Tamarkay	Climber
Swertia chirata	Gentianaceae	Chireto	Herb
Swertia pedicillata	Gentianaceae	Chireto	Herb
Taxus baccata	Taxaceae	Dhyangre salla	Tree
Thalictrum foliolosum	Ranunculaceae	Chitray	Herb
Tinospora cordifolia	Menispermaceae	Gurjo	Climber
Valeriana hardwickii	Valerianaceae	Nakkali	Herb
Viscum articulatum	Loranthaceae	jatamaasi	Herb
Woodfordia fruticosa	Lythraceae	Harchur	Shrub
Zanthoxylum oxyphyllum	Rutaceae	Dhangera Timmur	Shrub

in growth over the test period. Thus, it may be said that there is a potential that atmospheric CO_2 enhancement is likely to change the balance between the co-occurring species and thus affect biodiversity.

Enrichment of the atmosphere with soluble nitrogen compound has far more dramatic effects than CO_2 enrichment, which causes acidification of aquatic systems and favours grasses over forbs *(this is a particular group of plant)* (Korner, 2000)

Environmental pollution may lead to a potentially warmer climate, which may open the mountain zones to immigrant plants from lower elevational range and induce shift in the elevational range of many plant species. Thus a continuous climatic warming would be a biodiversity threat as it may narrow down the range of occurrence of certain plant species (Grabherr and Pauli, 1994).

Biodiversity and National Medicinal Plants Board

A medicinal plant board has been established under the Ministry of Health and Family Welfare (by a notification dated 24th November, 2000). The objectives of the board are to formulate strategies for conservation, cultivation, proper harvesting, processing, marketing and Research & Development related to medicinal plants (Pradhan et al., 2005). The board has identified 32 medicinal plants for commercial cultivation.

The National Medicinal Plants Board has been implementing a contractual farming scheme under which financial assistance to the tune of 30% of the total project cost (up to a maximum of Rs. 9.00 lakh) is provided to farmers/ growers for cultivation of identified medicinal plants. The scheme is operated through State Medicinal Plant Board. The Board has since revised the list of prioritized plants to include 80 plant species for cultivation and development. Of these, more than 60% plants are found in the Darjeeling Himalaya, indicating the richness of medicinal plant biodiversity in the region (Table-4).

Table 4
Medicinal plants available in Darjeeling Himalaya region as per the revised priority list of National Medicinal Plants Board

Sl. No.	Species	Common name	Sl. No.	Species	Common name
1.	Aconitum ferox Wall.	Vatsnabh	26.	Litsea glutinosa	Litsea
2.	Aconitum heterophyllum	Atees	27.	Mesua ferrea Linn.	Nagkeshar
3.	Aconitum palmatum D. Don.	Prativisa	28.	Mucuna pruriens Bak.	Kauso
4.	Acorus calamus	Vach	29.	Nardostachys jatamansi DC.	Jatamansi
5.	Asparagus racemosus Willd	Shatavari	30.	Ocimum sanctum L.	Tulsi
6.	Aloe vera Tourn ex Linn.	Ghritkumari	31.	Orchis latifolia Linn.	Panch aunlay
7.	Alstonia scholaris R. Br.	Satwin	32.	Oroxylum indicum Vent.	Totola
8.	Artemisia annua Linn.	Artemisia	33.	Panax pseudoginseng	Salanay
9.	Berberis aristata DC.	Daruharidra	34.	Phyllanthus amarus	Bhumi amlaki
10.	Centella asiatica Linn.	Mandookparni	35.	Picrorhiza kurroa	Kutki
11.	Cinnamomum zeylanicum	Dalchini	36.	Piper longum L.	Piplal
12.	Cinnamomum tamala	Tejpat	37.	Plantago ovata Forsk.	Isabgol
13.	Cinnamomum camphora	Kapur	38.	Podophyllum hexandrum	Papari
14.	Convolvulus microphyllus	Sankhpushpi	39.	Rauvolfia serpentina	Sarpgandha
15.	Digitalis purpurea Linn.	Foxglove	40.	Saraca asoca (Roxb.)de Willd.	Ashok
16.	Dioscorea floribunda	Dioscsorea	41.	Saussurea costus	Kustha
17.	Emblica officinalis Gaertn.	Amla	42.	Saussurea lappa C. B. Clarke	Kuth
18.	Ferula foetida Regel	Hing	43.	Solanum nigrum L.	Makoy
19.	Gloriosa superba L.	Bikh phool	44.	Swertia chirata Buch-Ham	Chireto

20.	Gentiana Kurroo Royle	Trayamana	45.	Taxus baccata Linn.	Dhayngre salla
21.	Ginkgo biloba	Ginkgo	46.	Terminalia belerica Roxb.	Barra
22.	Gmelina arborea Linn.	Gambhari	47.	Terminalia chebula Retz.	Harra
23.	Hedychium spicatum Ham.	Kapur kachari	48.	Tinosporia cordifolia Miers.	Gurjoe
24.	Hippophae rhamnoides Linn.	Seabuckthorn	49.	Withania somnifera (L.) Dunal.	Ashwagandha
25.	Holarrhena antidysentrica Wall	Kurchi	50.	Woodfordia fruticosa Lurz.	Dhangera

Biodiversity and CAMP Workshop

Rapid decline in the wild population of medicinal plants is a foregone conclusion. In order to come up with any plans of rehabilitation of medicinal plants it is imperative that a proper assessment of its population, the threat faced by it along with its conservation status is properly studied in a particular area. For suggesting management interventions and sustainable utilization of prioritized medicinal plants a method has been developed over the past 15 years. The assessment study is carried out through an intensive workshop known as Conservation Assessment and Management Prioritization (CAMP) process. The objective of the CAMP workshop is to provide strategic guidance for application of intensive management and information collection techniques to threatened groups. They provide a rational and comprehensive means of assessing priorities for intensive management within the context of the broader conservation needs (Anon., 2007).

The CAMP process pools together a broad spectrum of experts from among the scientists, foresters, wild life managers, ecologists, taxonomists, botanists, cultivators etc. for assessing the conservation status of selected plants. This process provides an opportunity to assign new IUCN Red List categories to the selected species. CAMP assessment could be done for a particular region, a state or even for a country.

In India, the first CAMP workshop for plant species was held at Bangalore for medicinal plant species of southern India in February 1995. This was followed by other workshops in 1996 and 1997. In 2007, a CAMP workshop was organized by the Foundation for Revitalization of Local Health Traditions (FRLHT) and Research Circle, Directorate of Forests, Government of West Bengal for the state of West Bengal. This workshop assessed and prioritized 46 different medicinal plants and assigned them Red List status as per IUCN guidelines (Anon., 1999). Of these 33 species of medicinal plants are found in Darjeeling Himalayas

(Table-5). This is not surprising, because species diversity in the hills is many times denser than in the plains and especially since Darjeeling Himalaya is an important component of biodiversity hot-spot of the Eastern Himalaya (Chhetri et al., 2007).

Table 5
Medicinal plants prioritized and assessed during CAMP workshop, held from 4-7 December, 2007 at Kolkata. Plant names prefixed with asterisk (*) are found in Darjeeling Himalayas

Botanical name	*Local Name*	*Family*	*Parts used*	*Threat status*
1	*2*	*3*	*4*	*5*
Abelmoschus moschatus	Mush-kdana	Malvaceae	Seed	NT
**Aconitum bisma*	Bikhuma	Ranunculaceae	Root	EN
**Aconitum ferox*	Vatsnabh	Ranunculaceae	Root	EN
**Aconitum spicatum*	Bikh	Ranunculaceae	Root	EN
**Alpinia calcarata*	Toroni	Zingiberaceae	Rhizome	EN
**Ampelocissus barbata*	Jarila lahara	Vitaceae	Stem	EN
**Aphanomixis polystachya*	Pittaraj	Meliaceae	Stem	LC
Aristolochia indica	Isharmul	Aristolo-chiaceae	Leaf, root	VU
**Asparagus racemosus*	Satavari	Liliaceae	Root	EN
**Berberis aristata*	Chutro	Berberidaceae	Root, bark	VU

(Contd...)

1	2	3	4	5
**Celastrus paniculatus*	Mulkagni	Celastraceae	Seed, bark	EN
**Cinnamomum bejolghota*	Bhale sinkauli	Lauraceae	Bark, leaf	VU
**Cinnamomum cecidodaphne*	Malaigiri	Lauraceae	Wood, Seed	EN
Desmodium motorium	Ban Chandal	Fabaceae	Root	VU
**Dioscorea prazeri*	Kukur tarul	Dioscoreaceae	Tuber, bulbil	EN
**Drosera burmanii*	Surya sishir	Droseraceae	Whole plant	EN
**Gloriosa superba*	Bikh phool	Liliaceae	Tuber	VU
Gymnema sylvestre	Gurmar	Asclepiadaceae	Whole plant	VU
**Gynocardia odorata*	Gantay	Flacourtiaceae	Seed	EN
**Helminthostachys zeylanicus*	Ekbir	Ophioglossaceae	Whole plant	EN
Ipomoea mauritiana	Bhumi-kumra	Convolvulaceae	Tuber, root	NT
**Litsea glutinosa*	Kukur chita	Lauraceae	Leaf,fruit,root	LC
Lumnitzera racemosa	Kripa	Combretaceae	Leaf, bark	VU
**Lycopodiella cernua*	Nagbeli	Lycopodiaceae	Whole plant	EN
**Mesua ferrea*	Nagkesar	Clusiaceae	Bark, flower	EN
Morinda citrifolia	Noni	Rubiaceae	Whole plant	VU
**Mucuna pruriens*	Kauso	Fabaceae	Fruit, seed	EN
Nypa fruticans	Golpaata	Arecaceae	Fruit, leaf	VU
**Olax nana*	Bhadu	Olacaceae	Leaf, fruit	VU

**Ophioglossum reticulatum*	Jibray	Ophioglossaceae	Tuber	EN
**Panax pseudoginseng*	Salanay	Araliaceae	Rhizome	CR
**Pericampylus glaucus*	Pipal patay	Menispermaceae	Root tuber	VU
**Persea glaucescens*	Kawla	Lauraceae	Bark, wood	CR
**Picrorhiza kurroa*	Kutki	Scrophulariaceae	Root	CR
**Podophyllum hexandrum*	Papari	Podophyllaceae	Whole plant	CR
Pterocarpus marsupiam	Bijaisal	Fabaceae	Bark, wood	EN
**Rauvolfia serpentina*	Sarpagandha	Apocynaceae	Root, leaf,seed	EN
Sonneratia caseolaris	Archa	Sonneratiaceae	Fruit, wood	EN
Stereospermum colais	Parania	Bignoniaceae	Bark	VU
**Swertia chirata*	Chireto	Gentianaceae	Whole plant	CR
**Taxus wallichiana*	Dhaygre salla	Taxaceae	Leaf, bark	CR
**Thalictrum foliolosum*	Chitre	Ranunculaceae	Whole plant	VU
**Toona ciliata*	Tooni	Meliaceae	Seed, bark	VU
Tylophora indica	Anantmul	Asclepiadaceae	Root, leaf	NT
**Viscum articulatum*	Harchur	Viscaceae	Whole plant	LC
Xylocarpus granatum	Dhandul	Meliaceae	Wood	VU

Abbreviations: NT – Near threatened; EN – Endangered; VU – Vulnerable; LC – Least concerned; CR – Critically Endangered

From the CAMP workshop, it became quite clear that Darjeeling Himalayas not only possesses one of the highest biodiversity in terms of medicinal plants, but it is also home to the most threatened species. As per IUCN criteria the risk of extinction of a species may be depicted with threat status from the least risk to highest risk in the order: **LC>NT>VU>EN>CR**. Thus, Darjeeling Himalaya with more than 90% of seriously threatened medicinal plants including 18% Critically Endangered ones demands a serious attention for conservation from all stakeholders. The scenario for West Bengal too is not that good with 41% Endangered and 13% Critically Endangered medicinal plants as prioritized by the CAMP Work shop (Fig-2).

Fig.-2. **Threat status of prioritized medicinal plants of Darjeeling Himalayas (DH) vis-à-vis the same for the state of West Bengal (as per CAMP workshop, Kolkata, 2007)**

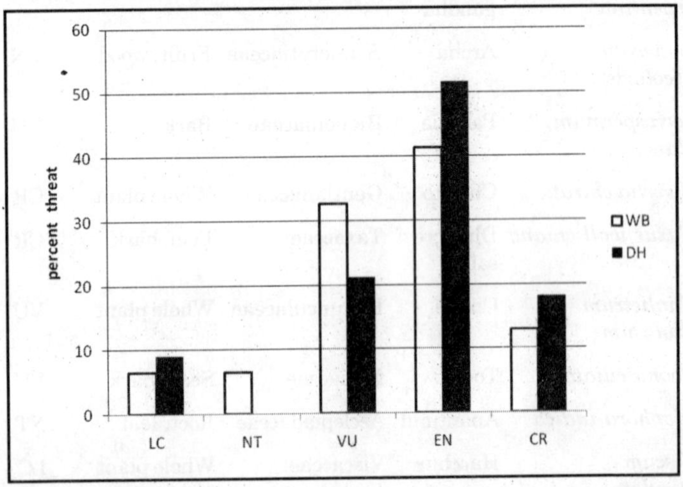

Conclusion

India is one of the few countries that is capable of producing most of the important plants used both in modern and traditional systems of medicine, due to the availability of wide variations of climate, soil, altitude and latitude. Most of the plants used by the drug industries, especially from high altitudes are harvested from the wild. This has led to the depletion of resources and extinction of species (Nautiyal and Purohit, 2000). Of the 814 threatened plant species in India, 113 taxa occur on Indian Himalaya (Nautiyal *et al.*, 2001). Increasing demand of medicinal plants from the Himalayan region has resulted in unsustainable harvesting and consequent loss of species. Besides, loss of forest cover due to developmental activities, natural calamities and demographic variation has led to the loss of biodiversity.

It has been realized that medicinal plants of the Indian Himalayan region offer an advantage in having much greater possibilities of providing new molecules used in drugs due to the extreme condition that they are subjected to. (Sharma et al, 1997). But a growing herbal market has posed a threat to biodiversity through over-harvesting that may lead to extinction of species (Chhetri, 2009).

The Darjeeling Himalayan region presents a unique and paradoxical position for having one of the most varied medicinal plant biodiversity, and at the same time also the most threatened medicinal plants in the country. Can we do something for conservation and sustainable utilization of this valuable resource? Involving the grassroots people, tribals, and cultivators seems to be the right way. Perhaps conservation initiatives should be linked to economy but it should always be kept in mind that the economy alone cannot justify the favoring of a few marketable plants at the cost of permanent loss of important biodiversity. The establishment of a dedicated research institute in Darjeeling, exclusively for medicinal plants, would be the right beginning for the development of this sector.

References

Anonymous. 2007. Conservation Assessment & Management Prioritization for Medicinal Plants of West Bengal. Department of Forests, Kolkata and Federation for Revitalization of Local Health Traditions, Bangalore.

Anonymous. 1999. IUCN/SSC Criteria Review Working Group. IUCN Red List Criteria review provisional report: draft of the proposed changes and recommendations. *Species* 31-32: 43-57.

Chhetri D.R., Parajuli S and Adhikari J. 2007. Antihepatopathic plants used by the Lepcha tribe of Sikkim and Darjeeling Himalayas, India. *Journal of Herbs, Spices and Medicinal Plants* 13 (3): 27-35.

Chhetri D.R. 2004. Medicinal plants used as antipyretic agents by the traditional healers of Darjeeling Himalayas. *Indian Journal of Traditional Knowledge* **3 (3):** 271-275.

Chhetri D.R., Basnet Deewa, Chiu Po Fong, Kalikotay S., Chhetri G and Parajuli Sippy. 2005b. Current status of ethnomedicinal plants in the Darjeeling Himalaya. *Current Science* 89 (2): 264-268.

Chhetri D.R.. 2005. Lepcha Ethnomedicinal Plants from Darjeeling Himalayas. In: Proceedings of *International conference on Promotion and Development of Botanicals with international co-ordination: exploring quality, safety, efficacy and regulations* (Mukherjee PK, ed.), pp. 234-239, Allied Book Agency, Kolkata.

Chhetri D.R.. 2009. Herbal Medicinal Culture of the Nepalis of Darjeeling and Sikkim. In: Indian Nepalis: Issues and Perspectives (Subba TB, Sinha AC and Nepal DR, eds.), Concept publishing Co., New Delhi, pp. 311-321.

Chhetri,D.R., Parajuli P., Subba G.C. 2005a. Antidiabetic plants used by Sikkim and Darjeeling Himalayan tribes, India. *Journal of Ethnopharmacology* 99: 199-202.

Grabherr G. and Pauli M.G.H. 1994. Climate effects on mountain plants. *Nature* 369: 448.

Hamilton L.S. 1987. What are the impacts of Himalayan deforestation on Ganges-Brahmaputra lowlands and delta? *Mountain Research and Development* 7: 256-263.

Hurni H. 1999. Sustainable management of natural resources in African and Asian mountains. *Ambio* 28: 382-337.

Kroner C. 1997. Alpine plant diversity a global survey and functional interpretations. In: *Arctic and Alpine biodiversity: patterns, causes and ecosystem consequences* (Chapin FS III and Korner C., eds.) *Ecological Studies* 113, Springer, Berlin, pp.45-62.

Kroner C. 2000. The alpine life zone under global change. *Gayana Bot.* (Chile), 57: 1-17.

Kroner C. 2004. Mountain biodiversity, its causes and function. *Ambio Spl. Rep.* 13: 11-17

Nautiyal B.P., Chauhan R.S., Purohit H and Nautiyal M.C. 2001. Assessment of germinability, productivity and cost benefit analysis of *Picrorhiza kurrooa* cultivated at lower altitudes. *Current Science*, 81 (5): 579 – 585.

Nautiyal M.C. and Purohit A.N. 2000. Cultivation of Himalayan Aconites under polyhouse conditions. *Current Science*, 78(9): 1062-1063.

Pradhan S., Lama P.C. and Chhetri D.R. 2005: Programmes on medicinal plants: the present scenario in Darjeeling and Sikkim Himalayas. In: *Natural resource based economy in the Eastern Himalaya* (D.R. Chhetri & P.C. Lama, eds.) pp. 69-86, Panchavati Greentech Research Society, Darjeeling.

Raghav, C. S. 2003. Medicinal plants on the touchstone of science. *Science Reporter*, 40 (2): 52-60.

Ramakrishnan, P.S. 1992. *Shifting agriculture and sustainable development: an interdisciplinary study from North Eastern India.* Carnforth: Paris & Parthenon Publishing Group.

Rameswararao B., Kesavulu M.M., and Apparao C. 2003. Evaluation of antidiabetic effect of *Momordica cymbalaria* fruit in alloxan diabetic rats. *Fitoterapia,* 74(1-2): 7-13.

Saini R.P. 2002. Medicinal plants of Darjeeling hills: A study by silviculture (Hills) division, Darjeeling. *The Indian Forester*, 126 (8): 822-837.

Sharma J.R., Mudgal V. and Hajra P.K. 1997. Review scope and perspectives. In: *Floristic diversity and conservation strategies in India. Vol-I.* (Mudgal V. and Hajra PK eds.), Calcutta: Botanical Survey of India.

Singh P.B. 2002. Medicinal plants of Kangra District in North West Himalaya. In: *Himalayan medicinal plants, potential and prospects* (Samant SS, Dhar U and Palni LMS eds.) Gyanodaya Prakashan, Nainital, pp. 185-196.

7

Understanding Social and Spatial Exclusion in India

*Sohel Firdos**

I
Social Exclusion

Introduction

The concept of social exclusion has made valuable contributions to the current understanding of poverty and socio-spatial polarization and segregation. The findings of a study commissioned by International Labour Organisation (ILO) on Social exclusion suggest that the concept of social exclusion has certain advantages. It provides an integrated and dynamic analytical perspective which reveals the processes; agency; and multi-dimensionality of disadvantage. It provides a framework for analysing the relationship between livelihood, well-being and rights. And by relating social and economic status to the notion of citizenship, it generates a collective moral responsibility for social integration. However, it is pertinent to mention here that

* Dr Sohel Firdos is grateful to Centre de Sciences and Humaines (CSH), New Delhi for providing financial support to write this paper.

despite wider application of social exclusion framework in understanding deprivation, there has been considerable debate in defining social exclusion. To Rodgers, it is basically about inadequacies with respect to certain means of living. "Social exclusion is seen in the growth of homelessness or urban slums, the declining hopes of the long-term unemployed, the lack of access to jobs and incomes of migrants and some ethnic minorities, the increasingly precarious nature of jobs on offer to new labour market entrants" (Rodgers, 1995).

De Haan (1995) finds that the potential of the approach for understanding poverty lies in the fact that present poverty studies tend to emphasize economic aspects and to pay less attention to political and cultural dimensions of poverty. The analysis of the inter-related processes of inclusion and exclusion could help one to understand the interplay between these dimensions. Nayak (1994) identifies the following types of exclusion: exclusion due to poverty, exclusion from employment and means to livelihood and caste based exclusion. "Social exclusion is a process which restricts the access of certain social groups to valued resources and entitlements, relegating them to the status of social outsiders. It is a multi-dimensional concept, conceived to capture different forms of social disadvantage — economic, social, political and cultural — that persist, in multiple variants and with different intensity, across nations. Consequently, in its widest connotation, the concept encompasses theories of poverty, inequality, racial/ethnic cleavages, long-term unemployment and citizenship." (Appasamy, Guhan, Hema, Majumdar and Vaidyanathan, 1995). Amartya Sen (2000) argues that social exclusion helps to understand poverty better and in a more comprehensive way. "While the underlying idea behind the concept of social exclusion is not radically new, the growing literature on the subject has helped to enrich causal understanding and empirical analysis of certain aspects of poverty and deprivation. To be excluded from common facilities or benefits that others have can certainly be a significant handicap that impoverishes the lives that individuals can enjoy. No concept of poverty can be satisfactory if it does not take adequate note of the disadvantages that arise from being excluded from shared opportunities enjoyed by others".

Entitlement and Social Exclusion

A useful way of conceptualizing the system of economic and social exclusion would be to focus on the problem as being one of lack of entitlement of economic and social power amongst a large section of the population. Following the pioneering formulation of Sen [1981] in the context of famines, the notion of entitlement suggests that it is not a drop in total food production but its poor distribution that may lead to famines. In fact there may well be famines in years of normal or even above normal food grain production if the distribution of purchasing power or entitlement is so skewed as to leave a large segment of the population below the margin of subsistence. The notion of entitlement refers to the actual or effective empowerment of a person to trade his original endowment of labour power and other factor incomes for food and other basic necessities. Thus it is not the total production of basic necessities that would make a material difference to the inherent health or strength of a community so much as the distribution of this total amongst the population. Thus the concept of the entitlement lends itself rather appropriately to address the question of social and economic exclusion in a wider context.

There is a fundamental sense in which the notion of entitlement is rooted in the neo-classical tradition of economics. The ability of individual agents to purchase goods and services is defined by their endowments, in the form of their labour power and other factor incomes and the exchange opportunities that are available. If it is the original endowment, or the distribution of purchasing power that is skewed then, quite clearly, the resultant distribution of goods and services traded would also, in general, be skewed. In order to bring about an alteration in the distribution of the original distribution of factor incomes itself, one needs to look at a theory of social or group or class action that ensures a more equitable distribution in the first place or better terms of trade or both. It is this that is required to ensure an equitable distribution of economic and social empowerment amongst the widest sections of the population. One should, however, note here that entitlements are not just a question of purchasing power as there may be means of acquisition other than via market exchange.

Entitlements are essential because they enable individuals to gain command over goods and services, and particularly those that may be thought of as "basic" goods. Once they are armed with this, individuals are then able to enhance their capabilities to realize their full potential. A larger entitlement or purchasing power allows a wider range of choices to an individual. The phenomenon of social exclusion may then be understood as one where there are sections of the population whose entitlements are limited or negligible.

It may here be important to think of entitlements as not just economic entitlements, but also political and social entitlements. In a privilege-laden or feudal society, the latter forms of entitlements are denied to all except a small minority of elite groups. Thus it is only an equitable distribution of economic, social and political entitlement that may be said to have the ingredients of a just society.

Social exclusion manifests itself in diverse ways. All those who are poor are necessarily part of the socially excluded in a fundamental sense, though the latter constitutes a wider category. Unemployment is, however, one of the mechanisms of exclusion. Poverty is an end state which itself implies a lack of entitlement to the basics for survival. Unemployment is a mechanism by which an adult member of society is denied a means to his or her livelihood and hence is prevented from his or her entitlement to basic goods and services. Even if the ranks of the unemployed may be compensated by a system of doles so as to enable them to acquire the basic goods necessary for their sustenance, they would still constitute an important category of the socially excluded, for employment is a vital constituent of an individual's dignity and self respect in society.

One may identify several other important dimensions of exclusion. One may, for example, consider exclusion from education, exclusion from housing, exclusion from property ownership, exclusion from democratic participation, exclusion from access to health services, exclusion from public goods, to name but a few. In addition to these one may think in terms of gender-based exclusion, exclusion of the old and infirm, exclusion of widows, and exclusion of the physically handicapped. Each of these deals with a special form of exclusion and deserves exploration for its own sake.

The notion of exclusion may also be widened to incorporate the notion of rights. For example, one may examine the notion of exclusion from the right to free speech, religion or, in the case of the young, the right to have a secure childhood. Exclusion from these rights constitutes a fundamental violation of individual liberty.

Finally we may also look at a dimension of social exclusion that is of central importance in India. This has to do with exclusion on the basis of caste. This is in some sense the most potent form of exclusion because this sociological characteristic is fixed at birth and is hence completely inflexible. Though daunting in themselves, the problems of poverty, unemployment, illiteracy and poor access to health care may still be addressed by concerted public action by government as well as non-governmental organizations. Eradication of caste-based exclusion is, however, a much slower and a yet more daunting task.

The caste system is based on the premise of a rigid social hierarchy. The traditional formulation of the fourfold division of the population into (i) Brahmins (ii) Kshatriyas (iii) Vaishyas and (iv) Shudras has mutated over centuries into a system of literally thousands of sub-castes, but what has remained in tact has been a system whereby it is the upper castes, comprising mainly Brahmins but also on occasion the other two castes, that have invariably been the dominant players in social and economic interaction while the Shudras have constituted the depressed classes of society. The large numbers of Harijans or untouchables are part of the Shudra caste in India. According to Ghurye (1969) the outstanding features of Hindu society are (i) segmental division of society, (ii) hierarchy, (iii) restrictions on feeding and social intercourse, (iv) civil and religions disabilities and privileges of the different sections, (v) lack of unrestricted choice of occupation, and (vi) restrictions on marriage, i.e. insistence on the practice of endogamy. Thus the structural basis of Hindu society is fundamentally based on the idea of exclusion. Caste is not merely a principle of social division but a comprehensive system of life dealing with food, marriage, education, association and worship.

Exclusion due to poverty

At a fundamental level, the concept of social exclusion must essentially be coterminant with poverty. This is especially true of a country like India where the depths to which the extent of poverty may go may not be quite comprehensible to those residing in the more affluent parts of the world. In the context of the more affluent countries, exclusion may refer to racial and ethnic minorities, immigrant communities, political dissidents, the old, single mothers etc. and, depending on the specific context of a particular country, these attributes may turn out to be rather significant determinants of social exclusion.

In the Indian context, however, poverty alone must be the most significant determinant of social exclusion. The question of poverty in the Indian context is, of course, related to the low level of per capita income and its highly skewed distribution. The literature on poverty in the Indian context is vast. A Working Group set up by the Government of India in 1962, of which some of the members were Prof. D.R. Gadgil, Dr. B.N. Ganguli, Dr. V.K.R.V. Rao and Shri Pitambar Pant, went into the question of the minimum standard of living and recommended that the national minimum for each household of 5 persons (4 adult consumption units) should be not less than Rs.100 per month in terms of 1960-61 prices or Rs.20 per capita. For urban areas this figures was to be placed higher at Rs. 25 per capita to cover the higher price of the physical volume of commodities on which the national minimum is calculated. This poverty norm of Rs.20 and 25, respectively, represented a broad judgement of minimum needs and was not strictly related to nutritional requirements, though the latter were taken into account.

We have described above the magnitude of poverty in India. The question arises as to how to combat it? In particular, it is of interest to know whether overall economic growth reduces poverty, i.e. whether a trickle-down mechanism is operative and also whether economic growth has been accompanied by increased inequality of income or consumption. This has been a matter of substantial discussion in the literature [Ahluwalia, 1978, 1986; Bardhan, 1970, 1973, 1974, 1986; Kakwani, 1980; Mellor & Desai, 1986; Minhas, 1970; Kakwani & Subbarao, 1990; Subbarao, 1985; Sen 1986; Srinivasan, 1986]. Some of the major conclusions of this literature are that trickle-down can, of course, occur but the process is seldom automatic. The beneficial effects of growth on the incidence of poverty can, but need not, be substantially offset by increase in the inequality of consumption. Anti-poverty interventions in terms of provisions of basic needs aimed at raising the income levels of the poor are essentially based on sound reasoning. It is seen repeatedly that growth and anti-poverty programmes ought to be seen not as substitutes but complements. The record on the health, education and nutrition fronts is so far unimpressive and public action in these areas needs to be bolstered substantially. There is some evidence to suggest that rural poverty is inversely related to the level of output per head of the rural population and positively related to the level of prices.

Exclusion from employment and means to livelihood

Next only to the incidence of poverty, one might argue that unemployment may characterize the phenomenon of social exclusion. Sen [1975] wrote three decades ago that "more has been written on the unemployment problem of India than on that of any other country in the world". The main difficulty with the writings on unemployment in the Indian context has been a lack of conceptual rigour in quantifying the extent of unemployment. Myrdal [1968] was of the view that the Western conception of unemployment is inapplicable in the context of India and he felt that one ought to develop a new "conceptual kit" to talk about the problem of unemployment in India. There are three main sources of data on unemployment in India, viz. (i) the decennial census (ii) the National Sample Survey, and (iii) the employment exchange registration and the Employment Market Information Programme [see Sen 1975, p. 118]. The 1961 census found only 1.4 million people unemployed in the whole of India, of whom 0.6 million were in rural areas, and it constituted less than 1 per cent of the rural labour force. This may on the face of it appear rather surprising and sanguine, but the puzzle may be explained by focusing on the criterion used to define employment. For seasonal work like cultivation, livestock, dairying, household industry, etc. if a person had some regular work for more than one hour a day throughout the greater part of the working season, he was taken as engaged in work. In the case of regular employment in any trade, profession, service, business or commerce, the basis of work would be satisfied if the person were employed during any of the 15 days preceding the day on which the investigator visited the household. Further, the census records as unemployed only those not working but seeking work. In other words they are based on the intersections of the class of unemployed in terms of the "income approach" and the class in terms of the "recognition approach" and are therefore narrower than each. Further, the "production approach" is not used. The approach of the NSS is somewhat narrower than that used in the census, but the basic approach of taking the intersections of the class of the unemployed in terms of the "income approach" and the "recognition approach" is retained.

An evaluation of employment [Papola, 1993] identifies the following broad trends: (1) Employment growth rate over the last two decades has been of the order of 2 per cent per annum whereas the growth rate of the labour force has been of the order of about 2.5 per cent. Employment growth decelerated in the early 1990s to 1.5 per cent per annum. (2) Employment has grown faster in construction, followed by manufacturing and services. Among the major sectors, agriculture has had the slowest employment growth over the entire period. (3) In manufacturing the unorganized sector has shown a relatively higher and the organized sector a relatively lower average growth rate of employment. (4) The recent deceleration in employment has been particularly sharp in agriculture and the manufacturing sectors. (5) Even though the employment structure has undergone some sectoral shifts away from agriculture and in favour of all other sectors, agriculture still continues to employ about two-thirds of the workforce. (6) The proportion of the self employed who constitute the majority of workers, has declined. There is a corresponding increase seen in the category of the casual wage earners. (7) Employment has grown much faster in the urban rather than the rural areas, but most of the growth has been in the informal sector. (8) In the rural areas, the non-agricultural sector has shown a significantly high employment growth, though it still contributes only about 22 per cent of the total employment.

In a study on the estimates of unemployment in India, Paul [1993] proposes a set of new measures to examine the temporal and regional variations in the level of unemployment in India. The new measure takes into account both the intensity and distribution aspect of unemployment. Using the new measure as well as the traditional measures of person rate of unemployment (PRU) and time rate of unemployment (TRU), Paul concludes, after examining NSS 32nd (1977-78) and 38th (1983) Round surveys that: (1) All the measures of unemployment except PRU show an increase in the level of unemployment in 1983 over 1977-78 in India. While the level of unemployment in the rural sector is lower than the urban sector, the contribution of the former towards total unemployment is very high. This is so obviously because a very high proportion of total labour force lives in the rural area. (2) The problem of underemployment is more serious than that of unemployment. About 18 per cent of the total labour force was found to be underemployed in 1983. (3) The level of unemployment varies from occupation to occupation. In the urban sector, the level of unemployment among the self-employed is lower than the others. In the rural sector the level of unemployment is the lowest among self-employed households and the highest among agriculture labour households. (4) Kerala shows the highest rate of unemployment followed by Tamil Nadu and West Bengal. The level of unemployment is the lowest in Madhya Pradesh, followed by Himachal Pradesh, Rajasthan and Uttar Pradesh.

The labour market in India is highly segmented [Kannan, 1994; Deshpande, 1983]. Among the key bases of this process of segmentation are caste, community and gender. Persons belonging to the lower social classes, e.g. scheduled castes, find themselves confined to a limited number of occupational categories, and this is so despite affirmative action policies in the labour market. Gender-based discrimination is evident across all categories of employment. It does not need belabouring that caste- and gender-based discrimination inhibits growth as well as distributive justice, in addition to directly contributing to the process of social exclusion.

Despite the conceptual difficulties in accurately measuring the extent of unemployment in India, it is clear that the magnitude of unemployment and underemployment is very substantial in India. These groups comprise an immediate an obvious category of the socially excluded. There appears to be a growing casualization of the rural workforce, with both men and women increasingly relying upon insecure and temporary employment rather than self employment or regular jobs [Ghosh, 1993]. Poverty is an important factor for increased female involvement in non-household productive work or entry into the labour market. There is a strong gender-based inequality in operation here, since the ratio of female to male wage rates, for equivalent work, is quite low.

Caste-based exclusion

We have mentioned above that poverty and unemployment may be deemed to be the most significant and fundamental categories of social exclusion. These are also the aspects that are emphasized by economists, perhaps to an obsessive extent, and to the exclusion of other equally significant and vital categories. One might for example focus on child labour or bonded labour, or one might like to look at slum dwellers or the category of destitutes or mendicants. While each of the latter categories would have a uniqueness of its own as a form of social exclusion they would ultimately form part of the broad categories of poverty and unemployment. Focusing on these two generic categories has therefore been the typical approach of economists. However, a separate examination of each of the other categories enables one to look at the nuances of other forms of exclusion. Thus while the group of child labour is without a doubt part of the poverty population, examining this as a separate category is instructive because they present important insights into the magnitude and the unique processes at work that give rise to these forms of exclusion.

Each of the above categories denotes an "end state" of the process of development and it is this end state that we regard as denoting a form of exclusion. This is true of poverty and unemployment, but this is also true of categories like child labour. It is not the case that these are end states that are necessarily fixed or rigid, though the chances are that a person who is in the midst of abject poverty is in all likelihood likely to continue that way unless something dramatic happens to change that status. Similarly the probability is overwhelmingly large that a child labourer, after he reaches adulthood, will simply be categorized as one of the poor, though, of course, some rags to riches stories are not totally absent in the country of such a vast population where the range of possibilities for human beings at any station in their lives must be phenomenally large. Thus, while one might say that each of these categories are virtually fixed for all practical purposes there is, of course, an outside chance that a bonded labourer may emerge into freedom through some constellation of probable or improbable happenings.

There is, however, one category of exclusion, vital in the Indian context among the Hindus, that is fixed at birth, and that is caste. This is a category that sociologists typically study, and economists tend to ignore, but the importance of this category can hardly be overemphasized. There are several aspects of exclusion that cannot be characterized by a person when he is described as being simply poor or unemployed. For example untouchables in many parts of India are prevented entry into certain temples, there are certain villages where members of the Scheduled Caste (SC) are forbidden use of certain village wells that are the exclusive preserve of the upper caste members. This is because to be in contact with a scheduled caste person is to be polluted.

Srinivas [1952] divides Hindu society into a fivefold categorization with the untouchables added on to the usually mentioned four castes, viz., the Brahmins, Kshatriyas, Vaishyas and Shudras. In some other descriptions untouchables form part of the Shudra caste. The idea of caste as a fourfold or fivefold division of society represents a gross oversimplification of facts. The real unit of the caste system is not one of the four or five *varnas* but *jati* (sub-caste), which is a very small indigenous group practising a traditional occupation and enjoying a certain amount of cultural, ritual and juridical autonomy. There are innumerable *jatis* or sub-castes. Professor Ghurye had estimated that there are some 2000 subcastes or *jatis* in each linguistic area, and there are 14 major languages in the country. Now, the attempt to fit the *jatis* of any region into the four or fivefold category is a very difficult affair. It is possible everywhere to say who are the Brahmins and the untouchables, but there is great confusion in the middle regions.

Membership of any *jati* or sub-caste may only be acquired by being born into it. The activities or occupation of several *jatis* are traditionally specified, in the sense that a cobbler's son can only be a cobbler and while there may be some other peripheral jobs he may do (for example, he may till land) he in any case cannot be a priest. Untouchables were forbidden to educate their children in village schools till very recently, and this may be so possibly even now, because the children of the higher caste Hindus would thereby be polluted. In social functions such as marriages, untouchables would typically eat separately from the other higher sub-caste groups. There could hardly be a more telling form of hierarchy and social exclusion anywhere else in the world.

Just in case we might have made too much of the fixity of *jati* or sub-caste, as distinct from other economic categories of exclusion which have some chance of flexibility, one should note that there is here, too, a notion of flexibility, but only in the next or subsequent births. According to the traditional Hindu view, birth is not an accident, but is preordained. The Hindu theological notions of *karma* (duty) and *dharma* (moral code) have contributed greatly to strengthening the idea of hierarchy which is inherent in the caste system. If a cobbler does his duty in all walks of life well and with devotion then in his next birth he may be born a Kshyatriya or even a Brahmin. By the same token, a bad Brahmin may be born a Shudra or an untouchable in his next birth. Thus as Srinivas puts it: "the caste hierarchy comes to be an index of the state of an individual's soul. It also represents certain milestones on the soul's journey to God". In the meantime, in the present birth, contact of any kind, touching, dining, sex and other relations between castes that are structurally distant results in the higher of the two castes being polluted.

This has led to spatial segregation of castes in a village in terms of their dwellings. The Brahmins as well as the untouchables usually reside in separate clusters of dwellings. The notion of social exclusion is therefore defined in a rather comprehensive and codified way. A Brahmin may not do the job of a cobbler or washerman, nor may an untouchable read the vedas in a temple. As regards the notion of pollution we may look at an interesting account given by A. Aiyappan in his study on the Nayadis in 1937 [see Srinivas, 1952: "A

Nayar must keep 7 feet from a Nambudri Brahmin, an Iravan must keep 32, a Cheruman 64 and a Nayadi 74 to 124. The respective distances between these lower castes are calculated by a simple process of subtraction: the Iravan must keep 25 feet from the Nayar and Cheruman 32 feet from the Iravan." This account would merely be comic if it were not the case that some of the above mentioned notions are still prevalent though perhaps in a diluted form in certain quarters of India.

In recent years there has been a resurgence of social and political consciousness amongst the lower castes generally and scheduled castes in particular. The term that is now commonly used is *dalits*, or the oppressed and exploited. The Constitution of the country had sanctioned reservations in jobs, educational opportunities and certain elected political offices for the scheduled castes (SCs) and scheduled tribes (STs), by way of positive discrimination. In recent years, however, the other backward castes (OBCs) have clamoured for reservations on the same grounds as jobs were reserved for SCs and STs earlier. The OBCs have now come to wield a substantial amount of political clout, particularly in certain State legislates (e.g. Uttar Pradesh) under the political leadership of the Janata Dal (JD) the Bahujan Samaj Party (BSP) and a number of other significant political parties of a leftist persuasion.

In the last decade there has been a raging debate about the logic and the desirability of the policy of social engineering via positive discrimination. Béteille [1991] is of the view that the system of reservation on behalf of disadvantaged groups in the name of distributive justice is likely to weaken the structure of institutions, and ultimately the structure of the State and civil society itself. Yet it is no doubt the case that after centuries of rigid hierarchies in the social and economic lives of people the new upsurge in the aspirations of the hitherto socially excluded classes is bound to take form in a set of social and economic policies that may not be fully rational. This would particularly be so when viewed from the vantage point of those who have been members of the creamy layer for several generations.

Shah [1991] is somewhat more sympathetic to social engineering via affirmative action. He is of the view that social backwardness or exclusion is not insurmountable. The process, however, is not automatic. It needs to be facilitated by interventions from within and without the political system. The policy of reservations is one such intervention that may provide temporary respite to the oppressed sections. But he feels that it has only a limited potential for bringing about such a change. He is worried that the legitimate demands of the oppressed are hijacked by their elite classes who are primarily interested in serving their own political interests. To them caste sentiments and solidarity are a means to political power. Yet the overall assessment of Shah is that in order to break the centuries-old system of rigid hierarchies, the system of reservation would help the process by now making it possible for larger number of families of different backward castes to enter the middle class. This would give rise to a process of genuine secularization of the middle class. Once a member of a family holds a white-collar job, that family should cease to get the benefit of reservation.

The electoral success of the political parties representing the backward castes in Uttar Pradesh (U.P.), one of the largest states in north India, is a pointer to the charting of a new political agenda in the country. For the first time in centuries the *dalits* or the underclass are beginning to count as a major political force. As the *dalits* are beginning to fight for their rights in various parts of the country, there are almost daily reports of caste fights and even caste wars. With growth of literacy and awareness, the oppressed classes are now standing up for their rights rather than taking things lying down as they always did for centuries. In public discourse people like B.R. Ambedkar and his mentor Jyotirao Phule (1827-1890) are now emerging as major new heroes. Phule, who was himself of the Shudra caste, had founded the Satyasodhak Samaj in 1873. The Samaj taught that all men and women were God's children and that one's greatness does not depend on one's birth. Both Phule and later Ambedkar stood for a society based on the principles of liberty, equality and fraternity. They both felt that it is only by fighting Brahminism that one could remove social oppression or exclusion.

Exclusion from Social Rights

In India, one may observe the most complete delineation of social rights. A particular finding of a study on social exclusion in India is that many aspects of social protection are linked with regular wage employment, which is obtained by only a minority of the workforce. Despite obligations written into the Indian Constitution to provide (within the limits of economic capacity and development) for the right to work, to education and to public assistance in the case of unemployment, old age, sickness and disablement, entitlements have been concentrated on "organized workers" who number less than 10 per cent of the workforce, and implementation of social assistance schemes for other population groups has been patchy.

There is a complex set of forces, economic and political, which both enable and entrain the expansion and effective delivery of rights. Appasamy and others (1995) considered in some detail the factors underlying basic needs satisfaction in India. They reported that, underlying the expansion of the provision of public goods in the health and education sector, and also improvements in water supply and sanitation, is a general commitment to promote economic growth. This has been complemented with: specific State interventions to augment the asset base of poorer sections of the population and to provide employment; efforts to move towards more universal provision and higher standards of the services for which the government has accepted prime responsibility; and mandatory reverse discrimination and affirmative action in favour of Scheduled Castes and Scheduled Tribes facing historically accumulated disadvantages. The study argues that the progress which has been made in reducing exclusion from such basic needs reflects two major countervailing tendencies. On the one hand, mass electoral politics based on adult franchise has been important in increasing the political clout of the poor. On the other hand, the effectiveness of programmes targeted at the poor has been reduced by mis-targeting, leakages, corruption, and lack of accountability. Moreover, as special programmes have been implemented, additional groups have also begun to demand special treatment. The politics of caste have become more conflictual in recent years.

Some of the studies, particularly on Kolkata, show that people living in situations of extreme material poverty did not necessarily view themselves as "excluded" or "marginalized". It is suggested that people are so used to a hard life and material deprivation that they do not always see it.

Exclusion from goods and services

Poverty is usually seen in terms of low consumption levels, and measured in relation to a basket of goods and services, including some which may be provided through public channels such as education and health care. In this framework, exclusion from consumption is mainly a result of inadequacy of purchasing power (or of entitlements, in Sen's framework), and one needs to look further back, in the reasons for low purchasing power, to find the principal mechanisms of exclusion. However, some aspects of consumption can be legitimately interpreted directly in terms of exclusion. The visible growth of mass consumption among affluent minorities clearly intensifies the sense of exclusion among other groups even if absolute poverty does not increase. As Wolfe expresses it, there is exclusion from the consumer society, from a high consumption way of life. Housing may also play a direct part in exclusion, as the clustering of the poor in areas without sanitation, good water, recreational facilities, reliable electricity and other services may directly promote exclusion — residence in a particular locality may stigmatize, may lead to poor health, may make community integration difficult. Spatial segregation has become a major concern in the French literature on exclusion, and one can see parallels in the literature on slums in India.

Basic needs: Access to housing, water supply and sanitation

Exclusion of disadvantaged groups from the Indian mainstream is most apparent in respect of housing. The proliferation of squatter settlements and slums in all the major urban centres of India and the segregated settlements of the Scheduled Caste/Scheduled Tribe populations of rural India are a visible reminder of their physical exclusion. Ironically, government policies and programmes have accepted the status-quo, namely, physical exclusion or clustering of the poor in separate settlements both in urban and rural areas. Since housing is largely a private sector activity, government policies have been restricted to a large extent to providing tenure, services, etc. Despite vigorous efforts by the National Campaign for Housing Rights, de facto government policies have not been able to guarantee decent housing as a "social right". Data collected by the National Sample Survey and others only serve to confirm the depressing state of affairs in this sector. Thirty-nine per cent of rural households and 11 per cent of urban households live in temporary structures — much of it in poor condition and unsuitable protection against rain, wind, etc., and 89 per cent of rural households and 31 per cent of urban households lack latrine facilities. Inadequate waste disposal facilities contribute to communicable diseases. Seventy-two per cent of households in urban areas supposedly have access to safe drinking water, but in rural areas the proportion is less than half. The outbreak of plague in the city of Surat, and the numerous epidemics of water-borne diseases such as cholera, gastro-enteritis, malaria, in the major metropolitan centres of India soon after the monsoon, clearly indicate the unsatisfactory state of affairs with regard to urban water supply and sanitation. The largest casualties are those who live in the shanty towns and squatter settlements. The poor are also the most vulnerable to industrial accidents, as was evident in the Bhopal gas disaster. Why does such exclusion take place? In rural areas, caste-based discrimination is an explanatory variable. There is not only segregation in housing, but also in terms of access to water supply. In many villages, the Scheduled Castes have separate water supply sources; while in villages where water is shared, they must either have the water poured out for them or draw water at different times of the day. Despite various attempts to end the sickening practice of "scavenging", it is carried on in many villages and towns because those involved in this occupation have no other means of livelihood.

In urban areas, the poor are priced out of the land and housing markets. Since the settlements are often "illegal", they are not provided with services such as water supply, sanitation and garbage disposal. Urban development programmes, such as slum upgrading, accept the physical clustering (except in hazardous areas), but seek to provide better services.

Despite all these shortcomings, these excluded groups have used various survival strategies. The Indian political system confers one major benefit to all, namely, the right to vote. The urban poor in particular have used this right to extract some benefits for themselves —land tenure, food, and some urban services. The Scheduled Caste population of rural areas has tended to migrate to urban areas not only in search of employment, but because they can merge into the urban "melting pot" and thereby avoid overt exclusion. But their poor economic status may still exclude them from full participation in the benefits that urban areas provide.

Exclusion from land

Exclusion from land is a critical issue in India, and widely associated with poverty and insecurity — but land is a source not only of livelihood but also of social integration in a broader sense. Historically, the factor of caste has operated as a major determinant of social status in many critical spheres of life. One such critical area is the access to land, the main productive asset in an agrarian society, such as ndia. Caste groups are surrogate classes and a division between those who owned land and those who do not can easily be perceived along caste lines. The upper castes are invariably the landed elites. On the other hand, those in the lower rungs of the caste society are largely landless, working on the land estates of the ownership farmers. It is understandable that in a vast country like India with diverse agro-climatic conditions there were regional differences in the types of landownership and the tenurial relations. It is also understandable that land systems of British India, such as those described as *ryotwari, mahalwari or zamindari*, came in response to the diverse agro-ecological conditions, and the regionally diverse socio-political structures. In the course of time, on the *zamindari* lands, an Indian form of feudalism emerged in which the landed elites were the high castes while the labour supply came from the low castes. Exceptions apart, there is therefore, a broader correspondence between the ritual hierarchy of caste and the distribution of ownership rights on land. The landed elites are the *dominant* castes. S.M. Mandavdhare (1986) has investigated the question of caste and land relations empirically by selecting a sample of fifty villages from the Marathawada region of Maharashtra. His data based on individual caste households show a broad conformity between the status of a caste-group and the proportionate share of land it holds. This study also suggests that in villages where the lower castes are politically organized, inter-caste inequality has considerably reduced. This shows that other things remaining the same, political mobilization may help the oppressed sections in their economic rehabilitation. State intervention by introducing land reforms to reorganize the ownership patterns of land has not helped much in reducing exclusion particularly in the states located in north India. Moreover, in recent times, a number of development initiatives of the different state governments have led to ejection of local residents in favour of manufacturing companies, industrialists, real estate developers. The case of Singur in West Bengal and SEZs in all parts of India are the glaring examples of such developments.

Exclusion from security

Security has a number of different dimensions. One is physical — security of the person, in terms of safety, freedom from risk of physical violence. To this might be added freedom from environmental insecurity. A second concerns security of livelihood. A third involves protection against contingencies — accident, ill health, death. This issue, which has been taken up strongly by UNDP in recent years (see, for instance, UNDP [1994]), is one to which the analysis of exclusion can be usefully applied. Insecurity of livelihood is linked to the exclusions from land and labour markets discussed above, but focuses on risk — risk of loss of land or of employment, and the possibilities for finding alternative income sources.

In comprehensive social security systems, unemployment insurance or social assistance programmes provide a safety net, but these are rare in India. Instead, informal sharing mechanisms at the community or family level, sometimes structured as pooled resources in mutual credit funds provide a — usually unsatisfactory — fall-back.

Protection against contingencies through social insurance is more widespread, but it is usually those who are least well placed to withstand the contingencies who are least well insured. Appasamy et al. highlight the ways in which inadequacies in social security coverage reflect and lead to exclusion in India (see also Guhan [1994]). They argue that this issue is underemphasized in policy design and that, with appropriate promotional, insurance and social assistance measures, an acceptable degree of security can be attained.

Exclusion from human rights

A major advantage of the notion of exclusion is that it permits one to consider both rights and welfare within a single framework. The attainment of a living standard adequate for health and well-being in fact forms part of the Universal Declaration of Human Rights but, in practice, inadequate consumption standards are usually analysed in welfare terms. Nevertheless, it makes a great deal of sense to link these two perspectives, for the attainment of particular basic human rights may well be a precondition for overcoming economic exclusion. Thus, the right to freedom of assembly and expression is surely important as a basis for effective mobilization and organization, in turn a precondition for overcoming other forms of exclusion. Where competing interests are at stake in the struggle against exclusion, equality before the law is also a powerful weapon in the hands of the less powerful.

The reverse relationship is also true, as economic and social exclusions affect the attainment of other rights. Extreme need leads to abuse and forces individuals to abandon rights to legal protection or freedom of association. The exclusion of minority groups from full participation in modern economies may render them vulnerable to exploitation and diminish their ability to defend their assets and their civil rights. These exclusions are interdependent with exclusions which are more economic in nature. For instance, the exclusion of populations from land may start with a process aimed at restricting other rights.

Exclusion and macro-economic development strategy

A central aspect of the analysis of exclusion is the idea that it is embedded in the way societies function. Thus, differences in development paths and in macro-economic and structural adjustment strategies, imply equally varied patterns of social exclusion. Similar aggregate economic growth records may lead to quite different patterns of income inequality, quite different patterns of distribution of the benefits of growth. In this the institutional arrangements which mediate between economic and social development are crucial. They may include or exclude, limit gains to a few sectors or groups or spread advantage widely. Institutionalists within the regulation school, such as Robert Boyer [1994], show how a mutually interdependent set of institutional mechanisms underlies any particular growth path, generating the economic incentives for investment and production at the same time as they provide the social underpinnings of growth by setting the terms on which different groups participate in and benefit from development. Such arrangements therefore determine the pattern of both inclusions and exclusions.

An important application of this line of thinking is to the analysis of the social costs of structural adjustment. The retreat of the State and the restructuring of production systems to adapt to a more market- and externally-oriented economy have generated new patterns of inclusion and exclusion, notably in terms of access to social rights, to employment or to the best jobs in the labour market. Analysing the pattern of exclusion as an outcome of the interaction between economic and institutional mechanisms may provide significant contributions to the design of adjustment strategies which are more inclusionary than those in the recent past.

II
Spatial Exclusion

Spatial Exclusion based on Caste

The village mirrors the intricacies of the caste-based Indian society. The ritual hierarchy of castes expresses in space in the residential patterning of different castes within a rural settlement in whatever region of India it is located. In fact, the social morphology of rural villages in India is a spatial surrogate of the ritual hierarchical social order based on caste. What happened was that the most favourable sites within the village settlement (abadi) were occupied by the cultivating caste, the dominant caste. The Brahman houses were located in their neighbourhood. Separated by lanes were the quarters in which the middle ranking castes lived. The outer periphery of this agglomerated settlement was occupied by other service castes that generally performed clean services. There was yet another periphery in which the unclean castes, or the untouchables, holding the lowest rank in caste hierarchy, were living. Depending on the relief as in Rajasthan, or if the settlements were perched on mounds as in parts of Uttar Pradesh, the central promontory was occupied by the main cultivating caste, say Rajputs, Jats or Ahirs. At the lower level were other middle ranking castes and the low castes were situated on the outer periphery which was generally a low-lying area. Thus an altitudinal zonation based on caste hierarchy characterized the rural settlement. This model was universal and represented a type in itself. If Muslims, e.g., Sheikhs, Julahas or Fakirs, were a component of rural population, they also followed a zonal distribution pattern similar to that of the Hindu castes.

Spate noted that an Indian village did not follow a definite layout plan. However, "within the seemingly chaotic agglomeration there is, as a rule, strong internal differentiation, that of the separate quarters for various castes". His study of Aminbhavi, a village situated to the north-northeastern of Dharwar in Karnataka, although randomly chosen, was a typical example of the way caste factor operated in the residential patterning of different communities in a village. Based on a pioneering survey undertaken by C.D. Deshpande and his students and with subsequent inputs from L.S. Bhat, Spate's study can serve as a model. The village population consisted of Lingayats, an agricultural caste of Karnataka, Muslims, Jains, and Brahmans. The members of each caste occupied a solid block of contiguous houses in a lane named after the caste. The village population also included Talwars, Harijans, Wadars, Shikalgars and washerman. On the periphery of village lived the low castes. The occupational structure as in other parts of India is caste based. For example Lingayats are mostly tenant farmers whereas Talwars and Harijans are generally agricultural labourers. Members of other castes are carpenters, smiths, cobblers, washermen, barbers etc.

The general emphasis on caste in south (India) takes social fragmentation, allied with spatial separation, to the extreme, segregating the the untouchable in outlying cheris or sub villages, sometimes located several hundred yards from the main villages of which they are service components. This is indeed a climax of geographical differentiation; apartheid *(Spate 1972)*.

Despite the changes in Aminbhavi, the residential patterning continues to be the same. While caste-based residential segregation is universal, the village in south (India) presents an extreme form of spatial segregation (Ahmad 2004).

Caste as a determining factor of spatial exclusion in urban areas

The residential housing in urban centers is a continuation of the rural model – separate quarters for different castes. Take, for example, the towns of western Uttar Pradesh, such as Mathura, aligarh, Bulandshahr, Meerut, Moradabad, Bareilly, Badaun and Firozabad. There are separate mohallas for different Hindu castes. And this pattern is valid for all regions of India. If there are muslims in these towns they live in separate mohallas, katras or bustees. In the towns of British India there were mixed localities as well in which Hindu castes lived in the midst of Muslims or vice versa. However, segregation has increased with time and in many North Indian towns different communities live in separate blocks.

In a traditional setup, there are separate residential quarters for Brahmans, Vaishyas, blacksmiths, weavers, washermen, tailors and other artisan castes(e.g., Brahmans live in *mohalla* Chauban in Firozabad). The low castes such as the sweepers and the Kanjars generally live on the periphery of the town. There is an association between *mohalla* name and the castes that live in them. This is true for Muslim *mohallas* as well. There are separate quarters for separate castes (e.g., *katra* Sonaram, *katra* Pathanan, *mohalla* Shisgaran in Firozabad, *mohalla* Churiwallan in old Delhi). With development these towns have grown into large cities with new outgrowths mushrooming on the periphery but the caste-based segragation is still intact and the community based segregation has further widened. The neighbourhoods in the Mughal town of Shahjahanabad (now Old Delhi) present yet another example of caste-based residential segregation. There are *mohallas* with occupational distinction, such as Ballimaran, Suiwalan, Churigaran, Kucha Chelan, Gali Rajan etc. After the partition of the country in 1947, Shahjanabad (Old Delhi) changed a lot but the traditional *mohalla* structure is still intact.

Understanding Social Inclusion

The academic debate on social exclusion has been relatively silent on its assumed corollary. There have been some notable contributions to a debate on inclusion (cf. in particular the essays in Askonas and Stewart, 2000), but this has not been closely integrated into wider debate on exclusion. It therefore remains the case that in the majority of the exclusion literature the nature and meaning of social inclusion is merely implied or asserted. Cameron (2006) explores the concept of social inclusion more critically and without starting from the assumptions implicit in the conception of social exclusion as local poverty and marginalisation. The reasons for doing so are partly derived simply from a frustration with the inadequacies of the ways in which social inclusion tends to be defined only negatively within exclusion literature – ie, 'as not social exclusion'. Beyond intellectual pedantry, however, how we conceive of social inclusion is of considerable importance. This is because, without an adequate understanding of what we mean by inclusion, all our attention is focused on the problems and deficits of those labeled 'excluded'. As a number of commentators have noted, the way the social exclusion debate has developed places an enormous responsibility on those considered to be excluded to resolve their own problems – they have been made responsible for their own condition as a form of social 'pathology' (Bauman, 1998; Procacci, 1999; Byrne, 1999; Cameron and Palan, 2004).

What I want to argue here is that the general failure to develop a critical understanding of the real and discursive *geographies* of social inclusion has contributed to this. By critically reproducing a strongly *asymmetrical* and *normative* spatialisation of the concepts of inclusion and exclusion derived from the policy debate, the social sciences (including geography) have at best failed to offer a robust analysis of the way these concepts are being deployed. At best they have served to legitimize the ongoing redistribution of risk and responsibility away from the 'mainstream' and onto the 'excluded'.

Inclusion defined by exclusion

Where a conceptualization of inclusion does appear in the social exclusion literature, it is often indirect. Frequently, for example, it appears in invocations of 'normal' social expectation/participation or, more commonly, 'mainstream' applied to various things that people are understood to be excluded from: labour market, economy, society, culture, citizenship, etc. the meaning and location of the mainstream is routinely taken to be self-evident. As this implies, social inclusion is most commonly defined only *negatively* – as whatever is *not socially excluded*. For this reason, *much of the discussion of social inclusion is conceptually dominated by exclusion - social exclusion is the datum point against which social inclusion is both empirically measured and conceptually defined* (Cameron 2006).

Negative definitions of inclusion can also be found in very different approaches to social exclusion. Various attempts have been made to develop statistical metrics for social exclusion, consisting of linking conventional economic data on income distribution and unemployment with 'social' data (the status of schedules castes and scheduled tribes.). Recently, a committee was constituted by the Government of India, headed by Justice Sachhar, to collect data and report on the socio-economic status of Muslims. These in turn have given rise to other attempts to define and map social exclusion with increasing specificity by academics and policy agencies at state and national level. These exercises in enumerating exclusion have been extremely important in providing different ways of visualizing the cumulative effects of poverty and marginalisation on particular places and population groups. As such their contribution to the understanding of poverty intensity, dynamics and distribution has been invaluable. They are of interest here, however, less because of what they reveal about the real distributions of poverty and marginalisation, but more because of what they imply about conceptualization of social inclusion.

For all their empirical rigour, such analysis continue to reproduce and reinforce certain assumptions about those not 'excluded'. In particular, the ways the data are combined and mapped provide compelling visual images of social inequality: particular places and even whole regions appearing to be 'excluded'. By corollary, this imagery also seems to provide maps of 'included' places and regions. I say 'seems to' because we must be very careful here about the way we interpret such data. For example, although there are undoubtedly places that exhibit extremes of poverty and marginalisation, one of the reasons that they appear so clearly on the maps of exclusion is that the data themselves are already packaged by spatial assumptions. Put simply, if we collect local data we will tend to produce a local story from them because the problems we are identifying automatically *seem* to be features of place. For example, measuring unemployment and deprivation among erstwhile jute mill workers in some of the slums in Kolkata, does not factor in the causal effects of replacement of jute bags with synthetic bags or the division of the country that lead to the bifurcation of jute producing areas from the jute mills – none of which are local to the problems they helped create.

References

Ahmad, Aijazuddin. 2004. *Social Geography*. Rawat Publications, Jaipur.

Appasamy, Paul, S. Guhan, R. Hema, Manabi Majumdar and A. Vaidyanathan. 1995. 'Social exclusion in Respect of Basic Needs in India', in Rodgers, Gerry, Charles Gore and Jose B. Figueiredo (Eds). *Social Exclusion: Rhetoric, Reality,*

Response. International Institute of Labour Studies: Geneva: pp. 237-252.

Cameron, Angus. 2006. Geographies of welfare and exclusion: social inclusion and exception. *Progress in Human Geography*, vol. 30(3), pp. 396 – 404.

Cameron, Angus. 2007. Geographies of welfare and exclusion: reconstituting the 'public'. *Progress in Human Geography*, vol. 30(3), pp. 396 – 404.

De Haan, Arjan. 2007. Diversity in patterns of urbanization: The long-duree of an industrial area in Calcutta. *Indian Cities in Transition*. Orient Longman. Pp. 390-411.

De Haan, Arjan. 1994. *Bibliographical Review on Social Exclusion in South Asia*. ILO.

Mahadevia, Darshini. 2007. A city with many borders: Beyond ghettoisation in Ahmedabad. In Annapurna Shaw (ed.). *Indian Cities in Transition*. Orient Longman. Pp. 341-389.

Mohan, John. 2002. 'Geographies of Welfare and Social exclusion: Dimensions, Consequences and Methods'. *Progress in Human Geography*, Vol. 26, no. 1: pp. 65-75.

Nayak, Pulin. 1994. *Economic Development and Social Exclusion in India*. ILO.

Rodgers, Gerry, Charles Gore and Jose B. Figueiredo. Eds., 1995. *Social Exclusion: Rhetoric, Reality, Response*. International Institute of Labour Studies: Geneva.

Sen, Amartya. 1973. "Poverty, inequality and unemployment: Some conceptual issues in measurement", *Economic and Political Weekly*, special number.

Sen, Amartya.1975. *Employment, technology and development*. Clarendon Press, Oxford.

Sen, Amartya. 1981. *Poverty and famines: An essay on entitlement and deprivation*. Oxford University Press, Delhi.

Sen, Amartya. 1983. "Equality of what"? in *Choice, welfare and measurement*. Oxford University Press, Delhi.

Sen, Amartya.1986. "Dharm Narain on poverty: Concepts and broader issues", in Mellor, J.; Desai, G. (eds.).

Sen, Amartya. 1990. "Justice: Means versus freedoms", *Philosophy and Public Affairs*.

Sen, Amartya. 1992. *Inequality reexamined*. Clarendon Press, Oxford.

Sen, Amartya. 2000. *Social Exclusion: Concept, Application, and Scrutiny*. Social Development Papers no. 1. Asian Development Bank.

Spate, O.H.K. and A.T.A. Learmonth. 1972. *India and Pakistan*.

Béteille, A. 1969. *Social inequality*. Penguin, Harmondsworth.

Béteille, A .1991. "Distributive justice and institutional well-being", *Economic and Political Weekly*, annual number.

Ghurye, G. S. 1969. *Caste and race in India*. Bombay, Popular Prakashan.

Papola, T. S. 1990. "Unemployment, poverty and inequality", in M. Adiseshiah (ed.): *Eighth Plan Perspectives*. Lancer International, New Delhi.

Ghurye, G. S. 1993. "Labour market monitoring for employment planning: A framework illustrated with the Indian experience", in Muqtada, M.; Hildeman, A. (eds.): *Labour Markets and Human Resource Planning in Asia: Perspectives and Evidence*. International Labour Organization.

Mandavdhare, S.M. 1986. *Spatial Patterns of Caste Distribution and Land Relations in Marathwada*. Unpublished Ph.D. thesis, Centre for the Study of Regional Development, Jawaharlal Nehru University, New Delhi.

Paul, Satya. 1993. "Unemployment in India: Temporal and regional variations", in *Economic and Political Weekly*, pp. 2407-2414.

Planning Commission. 1993. *Report of the expert group on estimation of proportion and number of poor*, Perspective Planning Division, Government of India, New Delhi.

Shah, Ghanshyam. 1991. "Social backwardness and politics of reservations", *Economic and Political Weekly*, annual number.

8

Migration from North Eastern Region to the Rest of the Country: Recent Patterns and Trends

Reimeingam Marchang

Migration along with mortality and fertility is considered to be one of the vital population processes in changing the size, structure, composition and distribution of population. Migration is the most difficult to measure as it is a fluid, continuous, non-discrete and poorly defined measure (Bernard D'Sami). It can either depopulate or overpopulate an area depending upon the level of economic activities and is an important process of urbanisation and social change. Its process affects the areas to which migrants have moved and areas which they have left. One important facet of study on population is the study of migration arising out of various social, cultural, economic, political or psychological reasons. Migration is increasingly taking place due to these various reasons from North Eastern Region (NER) to the Rest of the Country in the recent years. The mobility of a person has rapidly increased ever since the economic liberalisation, privatisation and globalisation particularly. Among the various reasons the booming of economic activities specifically in large cities like Delhi or Bangalore attracts most of the migrants. As Kundu (2007:353) also mentioned, "new employment opportunities are coming up in selective sectors and in a few regions/urban centres". In addition, the existence of good educational institutions also attracts the migrants from the region. In the process most of the migrants are migrating towards these areas that are comparatively more developed.

The objectives of the present study is to analyse the recent patterns and trends of migration and to dissect the various responsible reasons which act as a push as well as pull factors for migration from NER to the rest of the country over the years. It is important to study on migration because for a large country like India, the complexity of movement of population in different parts of the country helps in understanding the dynamics of the society. Further, at this crucial juncture of economic development in India in general and NER in particular a study on migration assumes special importance for further policy formulation and implication.

Socio-economic Factors for Migration

The increase of population and educational level in general and educated persons in particular with the limited supply or opening of jobs suitable with their educational attainment itself originated the problem of migrations from the region. Presumably, migrants from this region to the rest of the country particularly to the western or southern states of the country are mostly educated. Moreover, marginal expectation of migrants is always higher than the non-migrants. Further, the region is still underdeveloped and its economy is growing at moderate rate even in the era of globalization. During the first half of the Nineties "most of the economies of NER slowed down" (Sachdeva, 2000:21). This limits the creation of adequate employment avenues, which acted as a push factor for migration. Migration from the region to other parts of the Country is also a result of a rise in the attainment of higher education amidst the limited opening of job particularly in the organized public sector and a limited accommodation in higher education in the region.

NER covers 7.76 per cent (255,083 km^2) of India's geographical area. The region is one of the most ethnically and linguistically diverse regions in Asia with distinct cultures and traditions in each state. Due to this there is certainly a communication problem, which hampers the movement of various ethnic groups freely. In other words, the bonding of community is still close so exploration of different means of livelihood in various regions continues to be vague particularly in remote corners of the region. However, for the literate and educated it is not the case. These people's affinity for exploring new opportunities in order to enhance their quality of lives has no limit so they migrate.

According to census of 2001, the population of NER was 38.32 million, which comprises of only 3.73 per cent of the country's population. This is not proportionate with the contribution of NER in India's economy in terms of Net Domestic Product (Quick Estimates) that is only 2.88 per cent during 2000-01. Population of NER was growing at the level of national average of 1.94 per cent annually during 1991-2001. NE excluding the state of Assam, which are mostly tribal characteristic states, is growing faster in terms of population at an alarming rate of 2.44 per cent annually. This also contributes in the process of increasing out-migration from the region. In terms of population size the state of Assam has the largest share; while Mizoram has the lowest. Further, the latest census data also shows that majority of the population of Arunachal Pradesh, Meghalaya, Mizoram and Nagaland among the NE states were scheduled tribe; but for the states of Manipur, Tripura and Assam, majority of the population were non-tribal in 2001.

The population in NE is sparsely distributed however is increasing over the years. The density of population of NER was 151 persons per sq km in 2001, which has considerably increased from 124 persons per sq km in the previous decade. This has lead to an increase in agricultural density, urban growth, educated unemployment in urban areas, etc. in NER that has lead people to migrate outside the region. Meanwhile, the sex ratio of the population was 937 in 2001, a proportion that has improved from 925 in 1991. It indicates that there is not much gender discrimination so the possibility to migrate is same for both males and females. Further, literacy rates have considerably increased over the years. Literacy rates differential between males and females has narrowed down. However, it remained higher for males than females. It implies that the thirst for education is gradually quenching however not quenched because there is inadequate higher educational infrastructure, which leads to migration for higher education.

As per census of India data, agriculture and allied activities is the mainstay of the NE economy with about 63 per cent of the workers (main plus marginal) engaged in primary sector according to the census of 2001. Agricultural employment is declining due to the scarcity of agricultural land. It indicates that the growing population is pressed into agriculture, raising the density of agriculture. It also indicates that there is a wide prevalence of disguised employment. There tends to transfer labour from agriculture to non-agriculture sectors; however, non-agriculture sector is unable to absorb all the labour supply partly due to the growing educated people who mostly look for the formal/organised job. Moreover, this surplus labour are not absorbed in the non-agriculture sector which leads to seek job outside the region after collecting adequate information about the destination of migration. In 2001, employment in secondary sector is only 10 per cent and in tertiary sector is 27 per cent. Employment in non-agricultural sector is gradually increasing at the cost of declining employment in agriculture sector over the years. It calls forth for generating adequate job particularly in non-agriculture sector.

The corresponding sector-wise distribution of the NE economy (Net State Domestic Product) shows that primary sector contributed only 39 per cent (httpmospi.nic.in), which is too disproportionate compared to the employment ratio. This suggests the continuation of primitive farm practices with low production and productivity, which in turn suggest the presence of unexploited diverse natural resource endowments that are the potential for the development of the primary sector. The remaining shares of the economy come from secondary (15) and tertiary (46) sectors. In general, the contribution of primary and secondary sectors in the NSDP has declined as the contribution of tertiary sector increases over the periods for NER. It is clear that majority of the region's income is contributed by the non-agriculture sector, however, it alone is unable to generate as well as accommodate the growing demand for jobs resulting again to out-migrate from the region.

Concept and Definition of Migrants

Migrants are not required to be registered in India either at the place of origin or at the place of destination (Bhagat, 2005:3). In lack of registration of migrants, Census and NSS are the two main sources of migration data in India. Census provides data on migrants based on place of birth (POB) and place of last residence (POLR). Migrants defined on the basis of POB or POLR are called the lifetime migrants because the time of their move is not known (Visaria, 1980:2). Moreover, POB "migration data are not particularly useful indicator of trends in movement because they provide no information on timing of the movement" (Skeldon, 1986:761).

According to the Census of India "till 1961 census, migration data was presented with reference to place of birth only. The information on place of birth was being collected since 1872. In 1961 the scope of collecting information on migration was enlarged by including the rural or urban status of the place of birth and duration of residence at the place of residence. Since 1971 Census, data are being collected on the basis of place of last residence in addition to question on birth place". In 1981 census, the scope of enquiry on migration has been further widened by collecting information on 'reason for migration from place of last residence' in addition to the enquiry made in 1971 census. Thus a question on 'reason for migration' was introduced for the first time in 1981. "The pattern adopted in 1991 and 2001 Census remained same as in 1981 except that in 2001 Census, the rural urban status of place of birth was not collected" (Census of India, *Data Highlights*, 2001:6). The reasons for migration includes employment, education, family moved (moved with households in 2001), marriage, and others in 1981; two category such as 'business' and 'natural calamities like drought etc.' were included in 1991; however the category 'natural calamities' as one of the reasons for migration in 1991 was excluded and a new reason 'moved after birth' was added in 2001. The "others" includes all other reasons for migration not covered by work/employment, business (1991 and 2001), education, family moved, marriage, natural calamities (1991) and moved after birth (2001). This includes cases like "movement due to retirement, movement for economic reasons such as setting up of shops, starting of business, etc." in 1981 (Census of India, 1981, Series 1-India, Part-V A & B –I:27). However, since the census of 1991, "business" has been categorised as one separate reason for migration. Reason for migration provides useful insights for studying migration dynamics of population.

According to the Census of India, if the POB or POLR is different from the place of enumeration, a person is defined as a migrant. A person is considered as *migrant by POB* if the place in which the person is enumerated during the census is other than the person's POB. As a person could have migrated a number of times during his lifetime, migration by POB would not give a correct picture of the migration, taking place currently. A person, on the other hand, is considered as *migrant by POLR*, if the place in which the person is enumerated during the census is other than the person's place of immediate last residence. By capturing the latest of the migrations in cases where persons have migrated more than once, the concept of migrants by POLR would give a better picture of current migration scenario. At the time of enumeration in census, a person could have moved from another village or town in the same district, or from another district of the state, or another state in India or even from another country. Census provides migration data on all these migration streams by both the concepts to understand the dynamics in the movement of population and the broad reasons behind. Thus a person is considered as a migrant when the person is enumerated in census at a different place than the person's POB or POLR.

NSSO has been carrying out all-India household surveys once in five year in order to know the employment and unemployment situation and information on internal migration in the country. Being sample survey, the data have obvious limitations and are not helpful knowing the district level pattern in the internal migration within each state. The concept of Usual Place of Residence (UPR) is adopted by NSS to define migrants. A UPR is defined as a place (village/town) where the person had stayed continuously for a period of six months or more. According to NSS, a migrant is defined if the person had stayed continuously for at least six months or more in a place (village/town) other than the village/town where he or she was enumerated. The village/town where the person had stayed continuously for at least six months or more prior to moving to the place of enumeration was referred to as the last UPR of that migrated person (NSSO, 2001:14).

Analysis of Patterns and Trends of Migration

This paper analyses the changing patterns and trends of migrants (with all-duration of residence) from North-Eastern States/Region (as place of last residence) to the Rest of the Country (as place of enumeration) based on census data spreading the period from 1981 to 2001.

The general assumption that larger states in terms of population have larger proportion of migrants continues to be valid. Assam being the largest state in NE in terms of population with 69.57 per cent (Table 1) has the largest share of migrants with 68.64 per cent (Table 2) in 2001. Mizoram has the smallest population with 2.32 so it has the lowest share of migrants with 0.72 per cent. However, it is not the case for some states, for example Nagaland is the fifth largest state in terms of population size but is the second largest state in terms of the size of migrants with 19.15 per cent among the total migrants from NER to the rest of the country. The share of population is increasing in the tribal states over the years. It may probably be due to the lack of access of developmental infrastructure and health care information, tribal characteristics, illiteracy, poverty, and more importantly influx of migrants.

Table 1
Per centage ahare of population

NE States / India	*1981*	*1991*	*2001*
Arunachal Pradesh	2.55	2.74	2.87
Assam	72.94	71.05	69.57
Manipur	5.73	5.82	5.65
Meghalaya	5.39	5.63	6.05
Mizoram	1.99	2.19	2.32
Nagaland	3.12	3.83	5.19
Tripura	8.28	8.74	8.35
NER	100.00	100.00	100.00
	24801695	*31547314*	*38316918*
% of NER to India	3.73	3.76	3.73
India	*665287847*	*838567936*	*1028610328*

Note: Figures in italic are absolute. Figure of Assam is an interpolated figure for 1981. It is computed as the ratio between 'population of each NE States or Region' to 'total population of NER or India' in per centage term.

Source: Census of India 1981, 1991 and 2001.

Table 2
Per centage share of migrants from NER to the rest of the country

Last Residence	*1981*	*1991*	*2001*
Arunachal Pradesh	1.82	8.08	1.78
Assam	78.65	58.25	68.64
Manipur	4.59	6.60	3.21
Meghalaya	4.42	5.67	2.91
Mizoram	0.62	6.40	0.72
Nagaland	2.72	3.54	19.15
Tripura	7.18	11.47	3.60
NER	100.00	100.00	100.00
	232059	*346293*	*741509*

Note: Figures in italic are absolute. It is computed as the ratio between 'migrants from each NE States to the Rest of the Country' to 'migrants from NER to the Rest of the Country' in per centage term.

Source: Census of India 1981, 1991 and 2001.

A detail per centage share of "migrants from NE States/ Region to the Rest of the Country" to the "Population of NE States/Region" is given in Table 3. It is observed that the share of migrants from NER to the rest of the Country as a per centage to its population of NER was 1.94 per cent for persons, which is lower in comparison to the share of NER's population (3.73 per cent) to the country's population of approximately 1.03 billion in 2001. This indicates that NE people are not so mobile to the Rest of the Country because of a far distance factor. Apart from this other possible reasons are language problem along with the difficulties in socio-cultural adaptation and weak economic status in accessing the destination. The share of migrants-population was considerably lower for males than females in all the years particularly in 2001. It had increased throughout the periods for both males and females. A similar trend was observed for most of the NE states.

The increasing rate of migration is also contributed by the information and assistance given to the recent migrants by the previous migrants such as friends or relatives or the natives. In other words the increasing rate of migration is a condition of "chain-migration". Also the study of Michael Greenwood (1973) as cited in Mildred and Walter (1973:198), using 1961 census data for India found that "recent migrants do have a strong tendency to migrate to localities which had previously attracted natives of their region". Past migrant flows can be expected to influence current migration for several important reasons. Family and friends who have previously migrated from one region to another may provide information about their present location to persons residing in their former place of residence. Former migrants may also provide temporary food and shelter as well as ease social transition (Mildred and Walter, 1973:198). This likely takes place mostly to migrants from the rural areas. However, the rural to urban movement is the important flow for literate

Table 3
Migrants from NER to the rest of the country as a percentage to the population of NER

Last Residence	*Persons*			*Males*			*Females*		
	1981	*1991*	*2001*	*1981*	*1991*	*2001*	*1981*	*1991*	*2001*
Arunachal Pradesh	0.67	3.24	1.20	0.71	2.53	1.07	0.62	4.06	1.35
Assam	1.01	0.90	1.91	0.98	0.81	1.22	1.04	1.00	2.65
Manipur	0.75	1.24	1.10	0.88	0.96	1.11	0.62	1.54	1.09
Meghalaya	0.77	1.11	0.93	0.70	0.89	0.89	0.84	1.33	0.97
Mizoram	0.29	3.21	0.60	0.34	2.83	0.62	0.24	3.63	0.58
Nagaland	0.81	1.01	7.13	0.85	0.80	2.69	0.78	1.25	12.07
Tripura	0.81	1.44	0.83	0.85	1.19	0.86	0.78	1.71	0.81
NER	0.94	1.10	1.94	0.93	0.95	1.22	0.94	1.26	2.69

Note: Population of Assam for 1981 is an interpolated figure. Figure includes unclassified migrants. Rest of the country excludes J&K since it does not conduct census in 1991 for comparability.

Source: Census of India 1981, 1991 and 2001.

youth migrants (Sebastian, *Indian Youth: A Profile*, 1989:143). They form the majority of those who migrate in search of employment because of their energy, adaptability and aspirations.

Migration takes place not only on the motive of economic but also social, institutional and psychological factors. However, the basic cause of "voluntary migration is to achieve maximum individual satisfaction through obtaining better employment or wage or security or environment" (Santhapparaj, 1996:269). Hence "people migrate to maximise welfare" (Faggian and McCann, 2006:480) in terms of social security, economic security, etc.

This increasing trend is especially due to the lack of organised employment opportunities (where investment for the growth of industries and service sector is slow) that is the majority demand of employment among the educated unemployed, delivery of poor educational standards and most of the existing syllabus or course content inappropriate to apply in the local state of economic environment, non-availability of adequate seats in higher educational institution to curb an unemployment in the short-run, social insecurity due to the so called law and order problems which affects the psychology among the students which in turn influence in their future economic status.

Numerous studies (Jackson, 1969; Rossi, 1980; Friedlander and Roshier, 1966) in Britain have found that the "propensity to migrate increases with rising educational qualifications" (as cited in Cote, 1997:56). Generally, better educated people have more sophisticated and universalistic outlooks; have the financial resources to seek better opportunities elsewhere and are not constraint by local public housing policies which subsidise the poor but limit their mobility; are required to travel in order to obtain certain jobs; hold qualifications that are recognised and valued in all parts

Table 4
Migrants for employment, education and others as a percentage to the total migrants from NER to the rest of the country

Last Residence	Migration Stream	Reasons for Migration																	
		Work/employment						Education						Others					
		1981		1991		2001		1981		1991		2001		1981		1991		2001	
		M	F	M	F	M	F	M	F	M	F	M	F	M	F	M	F	M	F
1	2	3	4	5	6	7	8	9	10	11	12	13	14	15	16	17	18	19	20
Arunachal Pradesh	Rural-Rural	30.12	6.94	24.16	2.36	22.53	1.52	13.77	3.31	9.03	1.35	26.22	1.93	56.11	89.75	66.81	96.29	51.26	96.55
	Urban-Rural	20.68	6.39	16.53	3.66	17.92	4.28	6.14	9.40	9.20	8.42	20.68	6.89	73.18	84.21	74.27	87.92	61.40	88.83
	Rural-Urban	44.75	5.05	41.26	5.56	48.56	2.86	12.67	3.03	10.80	2.64	9.27	3.26	42.57	91.92	47.94	91.81	42.18	93.87
	Urban-Urban	34.91	4.02	36.21	5.77	34.02	3.61	14.53	3.59	7.61	4.75	15.27	7.75	50.55	92.39	56.18	89.48	50.71	88.64
Assam	Rural-Rural	13.75	1.36	13.15	1.66	22.94	0.90	2.34	1.04	2.27	0.60	1.42	0.09	83.92	97.61	84.59	97.74	75.65	99.01
	Urban-Rural	20.99	3.44	21.48	1.65	28.04	2.07	2.69	1.64	4.40	2.03	3.48	1.10	76.31	94.93	74.12	96.32	68.48	96.83
	Rural-Urban	35.30	3.98	30.61	3.45	44.94	4.10	6.43	2.77	5.67	1.68	5.04	0.88	58.28	93.25	63.72	94.86	50.02	95.02
	Urban-Urban	33.27	4.68	30.42	4.24	33.91	3.05	7.63	3.70	6.18	2.23	6.66	2.34	59.10	91.62	63.40	93.53	59.43	94.61
Manipur	Rural-Rural	34.68	4.47	20.93	5.66	30.46	1.32	1.84	1.40	8.45	1.92	19.16	3.15	63.48	94.13	70.63	92.42	50.38	95.53
	Urban-Rural	32.98	3.68	30.18	5.97	33.74	4.20	9.67	2.58	8.23	4.06	19.47	11.09	57.35	93.74	61.59	89.96	46.79	84.71
	Rural-Urban	45.85	4.20	39.53	5.63	40.27	6.48	21.62	6.01	13.06	8.34	27.45	15.67	32.54	89.79	47.41	86.03	32.29	77.85
	Urban-Urban	23.33	3.02	28.90	4.19	23.02	5.52	33.48	16.74	24.59	10.43	43.59	30.30	43.18	80.24	46.51	85.38	33.39	64.18
Meghalaya	Rural-Rural	19.83	8.18	20.57	2.89	26.29	1.60	3.10	3.61	7.48	1.65	8.06	1.68	77.07	88.21	71.95	95.46	65.65	96.72
	Urban-Rural	27.56	3.59	18.95	2.97	23.61	4.00	9.09	3.10	4.42	3.60	7.99	3.52	63.35	93.30	76.62	93.43	68.40	92.48
	Rural-Urban	43.86	6.38	38.15	5.87	53.01	4.68	7.78	4.54	10.50	3.15	5.46	2.23	48.37	89.08	51.36	90.98	41.53	93.09
	Urban-Urban	36.52	4.29	33.95	4.15	33.17	4.91	8.14	6.90	7.42	5.02	9.57	5.39	55.33	88.81	58.62	90.83	57.26	89.69

(Contd...)

1	2	3	4	5	6	7	8	9	10	11	12	13	14	15	16	17	18	19	20
Mizoram	Rural-Rural	58.02	0.00	12.34	3.48	35.74	2.40	6.17	0.00	4.66	0.85	5.64	2.08	35.80	100.00	83.00	95.67	58.62	95.51
	Urban-Rural	34.25	0.00	13.61	2.39	25.94	6.04	0.00	4.81	4.82	2.18	18.41	6.71	65.75	95.19	81.57	95.43	55.65	87.25
	Rural-Urban	22.22	6.67	37.09	4.42	52.96	6.53	17.90	5.71	8.33	2.43	17.75	9.20	59.88	87.62	54.59	93.15	29.29	84.27
	Urban-Urban	34.67	4.56	28.90	4.86	27.44	4.19	20.00	17.01	8.16	2.47	33.89	27.59	45.33	78.42	62.94	92.67	38.67	68.23
Nagaland	Rural-Rural	27.78	4.44	23.68	1.00	23.42	0.41	5.24	1.67	0.19	0.36	3.62	0.05	66.98	93.89	76.13	98.64	72.96	99.54
	Urban-Rural	24.97	4.68	23.33	1.29	22.63	1.66	3.35	2.90	10.10	4.53	6.54	2.18	71.68	92.43	66.56	94.17	70.82	96.16
	Rural-Urban	47.86	3.81	41.07	5.25	57.87	3.46	7.65	4.48	7.30	4.28	4.06	1.09	44.49	91.70	51.63	90.48	38.07	95.45
	Urban-Urban	41.49	6.30	36.51	2.87	37.43	3.46	12.51	9.39	14.80	5.21	10.83	5.03	46.00	84.31	48.68	91.92	51.74	91.51
Tripura	Rural-Rural	17.55	1.40	21.00	3.24	18.60	1.93	3.29	0.53	3.57	1.10	5.64	0.87	79.16	98.07	75.42	95.65	75.77	97.20
	Urban-Rural	23.38	2.35	25.75	2.33	23.81	2.30	4.68	4.23	4.45	2.70	5.34	2.40	71.94	93.43	69.79	94.96	70.85	95.31
	Rural-Urban	31.56	4.13	35.97	8.25	40.27	4.20	8.54	1.23	7.08	1.66	6.76	1.65	59.90	94.65	56.95	90.09	52.97	94.15
	Urban-Urban	34.52	4.78	32.94	4.32	35.57	3.56	8.27	5.38	10.64	3.25	11.14	3.32	57.20	89.85	56.43	92.43	53.30	93.12
NER	Rural-Rural	15.20	1.81	15.49	2.51	23.15	0.78	2.68	1.16	3.53	0.94	3.05	0.16	82.13	97.03	80.98	96.54	73.79	99.07
	Urban-Rural	22.11	3.45	21.31	2.14	26.98	2.27	3.40	2.06	5.16	2.78	5.98	1.85	74.49	94.49	73.54	95.08	67.03	95.88
	Rural-Urban	36.36	4.13	34.29	4.69	47.09	4.13	7.97	2.85	7.45	2.45	6.81	1.86	55.67	93.02	58.26	92.86	46.09	94.00
	Urban-Urban	33.09	4.60	31.27	4.29	33.60	3.37	10.20	5.03	8.37	3.22	10.09	4.30	56.71	90.37	60.36	92.49	56.31	92.33

Note: The figure of the rest of the country excludes J&K. Figures Exclude unclassified migrant. Due to rounding off the figure might not exactly sum up to 100. Movement for economic reasons such as setting up of shops, starting of business, etc. were included in the category 'others' in 1981; however, 'business' was separately categorised in 1991 and 2001. Thus, for comparability the category 'business' of 1991 and 2001 was club together in 'others'. Others in the present analysis includes marriage, move after birth, moved with household, natural calamities, movement due to retirement, movement for economic reasons such as setting up of shops, starting of business, etc. (business of 1991 and 2001), and all other reasons for migration not covered by work/employment and education. M – Males and F – Females.

Source: Census of India, 1981, 1991 and 2001.

of the country. Conversely, those who obtain only minimal school education may be expected to stay put because they have no special skills, which they could exploit elsewhere, and because of their better communal orientation – an attribute that is declared a working class value.

Further, the shares of migrants for employment and education from NER and from most of the NE states to the Rest of the Country was considerably higher for males than females for all the four streams of migration in 1981, 1991 and 2001 (Table 4). It is due to the economic dependency on male counterparts, male dominancy in the society, continuation of traditional concept of male as the bread earner, etc. Eventually, majority of the female migrants continues to be for other reasons than for employment and education. Further, a trend analysis showed that the shares of migrants for employment from rural NER to the rural Rest of the Country had increased during both 1981-91 and 1991-01 for males. It had declined during 1981-91, however increased during the latter period for urban-rural, rural-urban and urban-urban migrants. While for females, the shares of rural-rural and rural-urban migrants for employment had increased during 1981-91, however declined during the latter period. It had declined during 1981-91, however increased during the latter period for urban-rural migrants. And it had declined for migrants from urban to urban during both the periods.

Similarly, for migrants for education from NER to the Rest of the Country, the shares of rural-rural migrants had increased during 1981-91, however declined during 1991-01 for males. The shares of urban-rural migrants had increased; while the shares of rural-urban migrants had declined during both the periods. The shares of urban-urban migrants had declined during 1981-91, however increased during the latter period. And for females, the shares of migrants for education from rural-rural and rural-urban had declined during both the periods. The shares of urban-rural migrants had increased during 1981-91, however declined in the latter period. The shares of urban-urban migrants had declined during 1981-91, however increased during 1991-01. A similar trend follows for most of the NE states. The increasing trend is influences by both the pull factors like better employment opportunities and prospects, better educational infrastructures, non or minimal existence of social insecurity, better access to all sorts of information etc.; and the push factors like the slow economic development resulting to generate inadequate employment, social insecurity, too many "illegal taxmen" in many NE states so fear psychosis existed in the minds of many entrepreneurs who have capital, knowledge, skill, talent, etc. who can excel and play to contribute an accountable share in the state's income, inability to create and provide adequate seats in higher education while the base – i.e. up to secondary education – is wide, poor transportation and connectivity leading to slow access to information, etc.

A peculiar trend of migration for education prevails for the state of Manipur. For example, in 2001, it has considerably increased in comparison with other NE states for all the streams for females. Most importantly, among the NE states, Manipur had the largest share of female migrants for education in recent years. According to the latest census literacy rate of females of Manipur is close to 60 per cent which is the fifth highest among the NE states. This is an indication that the reason is not merely due to high literacy rate (compared to all-India level) but largely due to the existence of large number of educated apart from higher educational aspirations among the females of the state.

Furthermore, detail proportion of streams of migration from NE States/Region to the Rest of the Country is presented in Table 5. Most of the migrants from NER to the rest of the country were from urban-urban followed by rural-rural, rural-urban and urban-rural in 1981 and 1991; however, in 2001, majority were urban-urban migrants followed by rural-urban, rural-rural and urban-rural for males. While for females, the share was largest for rural-rural migrants followed by urban-urban, rural-urban and urban-rural in all the years.

Table 5
Proportion of streams of migration from NER to the rest of the country

Last Residence	*Migration Stream*	*1981* M	*1981* F	*1991* M	*1991* F	*2001* M	*2001* F
Arunachal Pradesh	Rural-Rural	32.99	35.84	29.31	54.42	24.07	44.06
	Urban-Rural	18.68	15.04	7.87	7.04	14.19	8.17
	Rural-Urban	21.44	22.39	41.04	24.69	26.47	19.12
	Urban-Urban	26.88	26.74	21.78	13.84	35.27	28.65
Assam	Rural-Rural	36.46	39.05	29.31	36.55	28.59	62.43
	Urban-Rural	16.54	14.87	12.06	11.38	8.09	5.01
	Rural-Urban	17.61	16.82	23.64	20.24	27.42	13.99
	Urban-Urban	29.38	29.26	34.99	31.83	35.90	18.57
Manipur	Rural-Rural	13.39	27.26	21.10	54.12	14.24	34.40
	Urban-Rural	12.39	12.97	8.38	6.15	8.77	5.41
	Rural-Urban	23.30	17.08	35.31	19.57	39.56	30.17
	Urban-Urban	50.92	42.68	35.21	20.16	37.42	30.01
Meghalaya	Rural-Rural	12.66	21.36	19.07	37.45	7.76	13.11
	Urban-Rural	15.37	11.50	14.25	9.78	11.02	9.79
	Rural-Urban	14.04	13.24	21.81	13.96	26.08	21.45
	Urban-Urban	57.93	53.90	44.88	38.80	55.14	55.65
Mizoram	Rural-Rural	19.17	16.97	35.03	51.05	12.13	26.72
	Urban-Rural	17.28	19.19	13.27	11.67	9.09	6.38
	Rural-Urban	19.17	19.37	25.23	17.42	39.85	32.12
	Urban-Urban	44.38	44.46	26.47	19.87	38.94	34.78

Nagaland	Rural-Rural	18.92	33.73	21.33	44.59	21.34	81.05
	Urban-Rural	25.98	16.83	19.70	13.39	4.94	2.12
	Rural-Urban	23.16	16.72	23.89	17.89	48.06	10.89
	Urban-Urban	31.93	32.72	35.08	24.13	25.67	5.95
Tripura	Rural-Rural	13.03	15.81	20.58	44.80	15.46	17.68
	Urban-Rural	9.93	11.82	8.95	8.48	7.87	8.57
	Rural-Urban	30.26	28.24	33.51	22.75	26.92	23.82
	Urban-Urban	46.78	44.13	36.95	23.96	49.76	49.93
NER	Rural-Rural	31.90	35.87	27.44	41.64	24.99	63.42
	Urban-Rural	16.12	14.50	11.64	10.27	8.08	4.60
	Rural-Urban	18.94	17.52	26.72	20.21	30.30	14.22
	Urban-Urban	33.05	32.12	34.20	27.87	36.63	17.76

Note: The figure of the rest of the country excludes J&K. Figures exclude unclassified migrant. Due to rounding off the figure might not exactly sum up to 100. It is computed as follows: 'Each stream of migrants' divided by 'total-migrants' from NE States/Region to the rest of Country multiplied by 100. Total migrant is the sum of migrants from rural-rural, rural-urban, urban-rural and urban-urban.

Source: Census of India, 1981, 1991 and 2001.

The share of rural-rural migrants from NER and most of the NE states to the rest of the country had declined during both 1981-91 and 1991-01 for males. In case of females, it had considerably increased for migrants from NER. A similar trend prevailed for most of the NE states particularly during 1981-91. The share of urban-rural migrants from NER and most of the NE states had declined during both the periods for both males and females. The share of rural-urban migrants from NER had increased during both the periods for males. In case of females, it had increased during 1981-91, however declined during the latter period for migrants from NER. It had also increased for migrants from most of the NE states during both the periods for both the sexes. The share of urban-urban migrants from NER had increased for males; while for females, it had declined during both the periods. It had declined during 1981-91; however, increased during the latter period for most of NE states for both the sexes. It can be explained by the lack of any employment opportunities at their origin of migration, the generation of rural jobs that are largely unskilled and manual in nature, increasing pressure on agricultural land, rising educational level, inability to established adequate capital-intensive industries for the unemployed, and generation of limited urban modern jobs that are mostly for the educated. All these factors in one way or other causes to unemployment which force people to migrate to areas where employment are likely to be available. These migrants adjust with any type of employment for a time being or even forced to take up whatever employment is available in order to sustain them and then in future after exploring and obtaining various experiences they seek and are employed in more stable/ permanent job associated with a higher wage. This is in line of what Todaro (1969:139) viewed labour migration in Less Developed Countries as a two-stage phenomenon. The first stage finds the unskilled rural workers migrating to an urban area and initially spending a certain period of time in the so-called "urban traditional" sector. The second stage is reached with the eventual attainment of a more permanent modern sector job.

Conclusion

Migration from NER is increasing over the years due to the rapidly growing population as well as growing educational level along with unavailability of employment avenues and inability to accommodate in higher education due to inadequate infrastructural facilities etc. which is all due to the prevalence of economic slack and buoyant economies in the region. The assumption that the state that has large/ small population size has large/small migrants is valid. A larger proportion of males continue to migrate for employment and education as compared to females. Urban to urban migration continues to dominate for males. In case of females, rural to rural migration is the largest. Lack of adequate educational infrastructure for higher education, underdevelopment, inability to generate adequate organised employment, educated unemployment, social insecurity, fear psychosis, etc. seemingly acted as a push factor for migration from the region. On other hand, better educational infrastructure, quicker access to information, availability of alternative jobs, rapid development particularly large cities, social security, etc. attracts migrants from the region to the rest of the country. Therefore, population should be controlled, development should take place, investment should move in particularly private sectors, labour intensive developmental work should increase, educational systems should be restructured, educational infrastructure should be made available adequately and socio-political problems should be solved in the region apart from others in order to reverse all these push factors for restructuring the population composition and to change the pattern and trend of migration from the region.

References

Bhagat, Ram B. (2005). 'Conceptual Issues in Measurement of Internal Migration in India'. XXVth IUSSP International Conference. Tours, France, July 2005.

Census of India. (1981, 1991 and 2001). Government of India.

Census of India (1981) Series 1-India, Part-V A & B –I. Government of India.

Census of India. (2001). Data Highlights. Government of India.

Cote, Guy L. (1997). 'Socio-economic Attainment, Regional Disparities, and Internal Migration'. *European Sociological Review*. 13(1).

D'Sami, Bernard, 'Migration Patterns and Challenges for Indians Seeking Work Abroad: A Special Focus on South India', *National Forum of Migrant Workers Rights*.

Faggian, Alessandra and McCann, Philip. (2006). 'Human Capital Flows and Regional Knowledge Assets: A Simultaneous

Equation Approach'. *Oxford Economic Paper 52*, Oxford University Press, httpmospi.nic.inrept%20_%20 pubnftest. asprept_id= nad03_1993_1994&type=NSSO.

Kundu, Amitabh. (2007). 'Mobility of Population' in Kaushik Basu (ed), *The Oxford Companion to Economics in India*. Oxford University Press, New Delhi.

Mildred, B. Levy and Walter J. Wadycki. (1973). 'The Influence of Family and Friends on Geographic Labor Mobility: An International Comparison'. *The Review of Economic and Statistics*, 55 (2).

National Sample Survey Organisation. (2001). *Migration in India*. Report No. 470, NSS 55th Round (July 1999 – June 2000), Ministry of Statistics & Programme Implementation, Government of India.

Sachdeva, Gulshan. (2000). *Economy of the North-East: Policy, Present Conditions and Future Possibilities*. Konark Publishers, New Delhi.

Santhapparaj, A. Solucis. (1996) 'Job Search and Earnings of Migrants in Urban Labour Market: A Study of Madurai Metropolis'. *The Indian Journal of Labour Economics*. 39 (2).

Nair, P.S., et al. (ed). 1989). *Indian Youth: A Profile*. Mittal Publications, Delhi, 1989.

Skeldon, Ronal. (1986). 'On Migration Patterns in India during the 1970s'. *Population and Development Review*. 12 (4).

Todaro, Michael P. (1969). 'A Model of Labor Migration and Urban Unemployment in Less Developed Countries'. *The American Economic Review*. 59(1).

Visaria, Pravin. (1980). 'The Level and Pattern of Economic Activity by Migration Status in India'. *Demography India*, 9 (1&2).

9

HIV/AIDS in Darjeeling: Historical Context and the Recent Scenario

Binu Sundas

Introduction

The HIV/AIDS epidemic is one of the greatest humanitarian crises of all times. It is destroying the lives and livelihoods of millions of people around the world. It causes death and untold misery, destroys families and communities. It breaks the social norms and values of the society (Nelkin et al. 1991) and it is a vicious cycle, which affects directly or indirectly every individual in a society. It has had tremendous impact on the development gains of several years and in a single stroke has ruptured the social fabric of the society. It is a source of enormous psychosocial, mental and physical stress to individuals who are infected as well as affected by it. That is why it has become a salient component in the agenda of health personnels across the world, and has received unprecedented attention. At present, there is no vaccine or drug to cure it and scholars like Kalichman (1998) think that even if a cure were to be found it would be so expensive that it would not be universally accessible. In the absence of vaccines and drugs to combat HIV/AIDS, a greater responsibility has been placed in the hands of human service professionals caring for People Living with HIV/AIDS (PLWHA) who has to confront many challenges resulting from HIV/AIDS infection – social, economic, lack of knowledge and shortage of trained medical staff.

The origin of HIV/AIDS has puzzled scientists as well as the public ever since the disease first came to light in the early 1980s. Since then it has been a subject of debates and a broad range of theories regarding its origin have emerged. These theories however do not converge to a common point but give diverse perspectives on the origin and spread of HIV/AIDS. It is unlikely that the truth regarding how, when and where HIV/AIDS actually originated will be ever known. A combination of many factors have attributed to the spread of the virus and undoubtedly certain features of the 20th century, like faster means of travel, communication and drug abuse etc have played a major role in its global spread.

Darjeeling is situated in a geographically strategic location. It shares international borders with Bhutan and Nepal. The borders of Bangladesh and China are also very close. In the year 1835, the Rajah of Sikkim ceded Darjeeling to the British East India Company (O'Malley 1907; Pinn 1990). Darjeeling as an urban center came into existence with the entry of the British in this hill station due to their need for a sanatorium and a summer capital (Lama and Chakraborty 2007). The British developed Darjeeling for reasons of its strategic importance and its potential to become a major trading center (Palit 2007). However, only with the introduction of tea and cinchona plantation did Darjeeling grow as an urban center (Pinn 1990). With the plantation of tea and cinchona the population of Darjeeling gradually started to grow from 100 in 1835 to 10,000 in 1859 and a staggering 1,73,342 in the early twentieth century (O'Malley 1907). With the growth of Darjeeling as a summer resort, it soon started to attract the people of the plains with its ample business opportunities. Darjeeling was soon flooded with migrants and business was taken over by the people of the plains. Darjeeling has today developed into a major tourist hotspot and has been connected with the larger global scenario. This exposure to the outside world has also had its negative fallouts. The youth are indulging in drug abuse and the past years have also witnessed large-scale migration among them. Due to the lower socio economic condition and a growing consumerist culture many are out to earn quick money by any means. Prostitution and drug abuse are on the rise. This situation has made Darjeeling highly vulnerable to HIV/AIDS.

Historical Context of HIV/AIDS in Darjeeling

Darjeeling district has witnessed political unrest for many decades now. Unresolved issues concerning identity, political separation from West Bengal, underdevelopment and lack of willingness on the part of the Central and State Governments to resolve these issues has brought about social and economic crisis in the district and particularly in the hilly regions. The Darjeeling Gorkha Hill Council (DGHC) was formed in 1988, after a long drawn out struggle for a separate state under the leadership of the Gorkha National Liberation Front (GNLF). However, the autonomous body was not given any real powers for development and employment generation. And there was misappropriation of any funds that came to the Council. This lack of vision and integrity on the part of the local leaderships plunged the district into deeper crisis. In 2008, the GNLF was unceremoniously overthrown by Gorkha Janmukhti Morcha (GJM), which has again spearheaded the campaign for a separate state. The autonomous Hill Council has no elected leaders and was declared defunct. Lately, the District Magistrate of Jalpaiguri district, with no local political participation, is looking after it. Years of political instability and step motherly treatment from the Central and State Governments have had tremendous socio-economic repercussions in the hills of Darjeeling. Infrastructural development remains abysmally low. Health facilities, roads, potable drinking water, educational facilities are acutely inadequate to serve the rapidly growing population.

In the context of HIV/AIDS it is important to note that the only three important hospitals (Victoria, Eden and Planters) were established during the British period. In the past 62 years, the government has not deemed it important to establish any new hospitals there. A few private hospitals have recently come up but they cater to a richer clientele. The sub-divisional hospitals, PHCs and the sub centers are ill equipped to handle grave illnesses. It was seen that among the hospital staff, there was not only a lack of correct awareness about HIV/AIDS, they also lacked empathy towards the positive people. This ill behaviour may have stemmed from the prejudices arising out of inadequate knowledge about the illness.

As far as education is concerned, barring the schools run by the missionaries, the government run schools lack infrastructure to produce well groomed students. Because of this many who pass out of these schools cannot compete for various state and national level competitive examinations and government jobs. There are also no good colleges in Darjeeling district so many students who can afford go to metros like Delhi, Kolkata, Bangaluru and Mumbai to study. There are no universities in the hilly regions of the district. The only university meant for the district is in Siliguri, where the medium of instruction is largely Bengali. The hill students therefore have to go elsewhere for higher studies. Being away from home at such a tender age, away from parental and societal control leads many to indulge in risk behaviours. So apart from those students whose parents can afford to send them outside Darjeeling to study, the rest have to make do with whatever is available. These youth are ill equipped to compete at various levels. They also do not have any vocational training. This has contributed to higher levels of unemployment. This is leading the youth to frustration, which in turn is leading to alcohol consumption and drug addiction. Under the intoxication of drugs and alcohol they are also indulging in risk behaviours. It is seen that even most of the students who have passed out from English medium schools have only managed to get employment in Business Process Outsourcing Offices (BPO). BPOs with its glitzy lifestyle, in a metro provide one with anonymity to pursue risk habits.

As far as employment in the hills is concerned, apart from the tea gardens that are facing closure due to an ever-competitive global market, there are no other industries. The hills of Darjeeling are very fertile and produce a variety of fruits and vegetables and cash crops but it is seen that, the hills, like the colonies during colonial rule only serve to supply raw materials. There are no food processing factories in the hills which could be established if the government desires.

The unpredictable situation of the tea gardens has exerted pressure on many to migrate in search for better livelihood. Many of these people who are unskilled labourers seek employment in the unorganized sectors and are very susceptible to diverse forms of exploitation. Many women come to the cities to work as domestic help. The nature of their jobs makes these people financially insecure and their long hours of work make it difficult for them to build social networks. In addition to such a financial position, the lack of social network so far away from home creates loneliness. So they seek association with other people and the easiest for them is to avail the services of Commercial Sex Workers (CSW). This makes them vulnerable to many diseases, including Sexually Transmitted Diseases (STD) and HIV/AIDS.

The youth who decides to stay at home become victims of frustration and depression. With no avenue to vent their anger these youth involve in the abuse of substances. Alcohol consumption and drug abuse among the youth of the hills of Darjeeling is common. At the time of fieldwork there were approximately 1200 youth enrolled at the different Indian Red Cross Society (IRCS) centers of the hills. In 2002 there were 114 alcohol related deaths in the Eden Hospital alone and in 2003 there were 815 enrolled on account of being alcohol dependent, at Kripa Foundation, a rehabilitation center for drug and alcohol dependents. This indulgence of the youth is also facilitated by the exposure to the outside world. Both these factors are associated with the spread of HIV/AIDS. The sharing of needles between the Intravenous Drug Users (IDU) and the engagement of these youth in risk behavior in the intoxicated state makes them vulnerable to HIV/AIDS. The fact that a large number of PLWHA in Darjeeling are former IDUs supports this fact. The location of the region, which shares international border with Nepal and is in close proximity with the North East states, which are the recognized routes for drug trafficking, also makes it easy for accessing drugs and this also adds fuel to an already volatile situation.

The exposure to the outside world and the lack of employment opportunities has changed the behaviour pattern of the youth in Darjeeling. The changing value system and the dilution of the traditional lifestyle have removed the taboo against sex. Girls have become more open to sex and the lack of economic pursuits has forced many into prostitution. Sexual practices among the youth are changing at a rapid pace due to various influences on their lifestyle.

The region is also poor in terms of the enormous health infrastructure required for the prevention and control of HIV/AIDS. During the time of conducting the field work there were no Antiretroviral (ART) centers, there was only one

Voluntary Counseling and Confidential Testing Center (VCCTC), which unfortunately was not accessible to the rural populace, there were no CD4 cell counting centers and the health personnel were also ill trained to provide care and support to the PLWHA. The web of all these complex variables has today made Darjeeling very vulnerable and susceptible to the spread of HIV/AIDS.

HIV/AIDS Scenario in Darjeeling Hills

Though West Bengal is considered a low prevalence state, it is highly vulnerable. West Bengal has been characterized as a low prevalence state because the median prevalence is below 1% for Anti natal Clinics (ANC) sites and below 5% in STD clinic sites. Presently a few hot spots indicate a raging concentrated epidemic in the state, which are Kolkata and its suburbs, and the urban areas of the adjoining districts of Howrah and Hoogly, North & South 24 Parganas and Haldia. The epidemic seems to be more significant among the lower socio-economic groups mostly comprising of migrants (WBSAP & CS 2006). Among the High Risk population West Bengal showed decreasing trend in all the groups other than MSM group where the prevalence rate increased from 0.54 in 2005 to 6.6 per cent in 2006. The HIV prevalence rate among the ANC cases remains somewhere at 0.4% over the years. HIV prevalence in STD in 2006 stood at 1.66% down from 2.79% in 2004 & 2.47% in 2005. HIV prevalence in Flying Sex Workers (FSW) has gone down from 6.8% to 6.6% & prevalence in IDUs decreased to 4.64% from 7.41%; only Men who have sex with Men (MSM) component has seen an upward trend in the 10th round of surveillance where prevalence rate shoot up from 0.54% to 6.12% (WBSAP & CS 2007)

In Darjeeling hills the prevalence of HIV/AIDS appears to be low, however, it is a major public health issue as the National AIDS Control Organisation (NACO) has recognized it as a very vulnerable area. The first case of HIV/AIDS in Darjeeling was identified in 2002, in an IDU after the setting up of the VCCTC at the Eden Hospital. Since then the number of people infected and affected has gradually increased. Till November 2007, the total number of HIV/AIDS cases had reached 141 in Darjeeling hills. The total number of men infected were 93 and the number of females infected were 43 and 5 were children. This rapid increase of the disease is also likely to be under reported partly because of the paucity of HIV/AIDS surveillance centers and the inadequacy of testing centers. Furthermore, the varying latency of HIV/AIDS among people suggests that some HIV/AIDS infected individuals may not manifest the symptoms for months or even years, even though they can transmit the virus to others. As one doctor said, *"The number is just a tip of an iceberg. The number can be ten folds more. Due to the lack of awareness and the inaccessibility of the VCCTC many are unable to get themselves tested. Again the well to do people does not get tested here as they fear the consequences of being positive. Therefore in all likelihood, the number of positive people in the hills can be much higher than what is presented by the VCCTC data."*

The demography of the reported HIV/AIDS cases in Darjeeling hills indicates that approximately two third of those infected are males. The infection is most prevalent in the 20-39 years age group of both sexes. The predominant mode of transmission is heterosexual sex, sharing of needles among IDUs and migration. A doctor from Kalimpong said, *"Here in the hills there are many reasons for the spread of the infections. People these days are engaging in casual sex and drugs addiction is also very high. The migrants are also responsible for the spread of this infection. Many join the army and bring the infection. In Kalimpong alone we have found 7 cases among the military people. Alcoholism is a major cause. As girls are easily available and after drinking people want to have sex. People do not have much knowledge about prevention of this infection. We can also say that the social changes taking place is also fuelling the spread of HIV/ AIDS."*

The HIV/AIDS cases have seen a gradual rise in the hills since the inception of VCCTC. In the first year, the number of cases detected was 7. Every year there was a rise in the number of cases and only in the 2004 was there a marginal decline, when only 17 cases were detected. The yearly figures from 2002 to 2007 are depicted in the graph below.

Fig.: **Yearly Infection in Darjeeling Hills- 2002-2007**

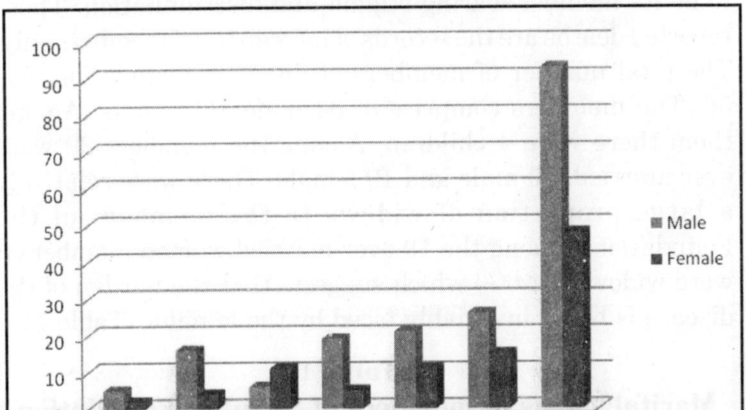

*Source:*Darjeeling VCCTC, Eden Hospital, Darjeeling, Aug 2007

Large number of PLWHAs was married and reportedly many are those who have been working outside Darjeeling. Till then 2276 tests have been done in the VCCTC and the positivity rate is 6.19 per cent. The prevalence rate is 0.03. The per cent of positivity among those coming to VCCTC suggests that if proper surveillance is done then the number can go very high as many people do not have access to VCCTC, as there is only one in the area. The VCCTC is also not equipped properly to handle the large number of cases that are being detected. As the hospital do not provide for any other services apart from testing and counseling most of the people who are found to be positive are asked to go to Shanker Foundation- a network of positive people, which maintains records of its members, for further counseling, care and support.

For this analysis members of Shanker Foundation have been taken into consideration. There have been 17 deaths reported by Shanker Foundation due to HIV/AIDS. However, all those who are tested are not its members. Many chose not to be its member fearing stigma and discrimination. These reported deaths are the records of its registered members only. The total number of members of Shanker Foundation was 56. The members comprise of 34 male, 22 female. Among them there were 4 children. Among the members 39 were ever married, 20 male and 19 female. There were relatively a large proportion of widows in the members of the Foundation. Among the 19 ever-married women members 6 were widows (31.6%) which suggests that the burden of the disease is being invariably faced by the females (Table 1).

Table 1
Marital status of members of Shanker Foundation

Marital Status	*Male*	*Female*	*Total*	*% Male*	*% Female*	*% Total*
Single	14	3	17	25.00	5.35	30.35
Married	19	13	32	33.92	23.21	57.13
Divorcee	1	0	1	1.78	0	1.78
Widow(er)	0	6	6	0	10.71	10.71
Total	34	22	56	60.70	39.27	99.97

Source: Shanker Foundation, Darjeeling, Aug 2007

The oldest among the male members of Shanker Foundation was 44 years old and the youngest was 5 years old and the oldest female was 39 years old and the youngest was 7 years old. The majority among both males and females belong to the 21-39 age groups. However, the educational background and income of the members was not available (Table 2).

Table 2
Age structure of the members of Shanker Foundation

Age Group	*Male*	*Female*	*Total*
5-20	3	1	4
21-25	6	6	12
26-30	5	4	9
31-35	9	5	14
36-40	8	6	14
41-45	3	0	3
46-50	0	0	0
Total	34	22	56

Source: Shanker Foundation, Darjeeling, Aug 2007.

Darjeeling has been identified as the most vulnerable to HIV/AIDS by WBSAP&CS. HIV/AIDS has already entered most of the high-risk groups and can spread to the general population as the practice of multiple sexual partners is common and people do not have a very comprehensive knowledge about HIV/AIDS, its prevention and treatment. The convener of one of the NGOs in Darjeeling succinctly described the risk of HIV/AIDS in the hills. He said, *"People do not have much idea about the infection and lack awareness and knowledge about it. Here the prevalence rate is 0.5-6.6. The prevalence of HIV due to sexual activities is 86 per cent and from blood transfusion is 2 per cent."*

It is clear that the advent of HIV/AIDS in Darjeeling will be a greater cause of worry for the government as the health infrastructure and the corresponding facilities are not sufficient to confront the epidemic. The socioeconomic conditions of the people are also not very good so that the people in the absence of any ART center can procure the medicine on their own and look after themselves. The spread of the epidemic will see the administration of the area ill prepared to tackle it.

PLWHA in Darjeeling Hills

Traditional approaches to the life cycle of human beings have tended to assume a fairly predictable progression through a sequence of life stages, from birth to death via childhood, marriage and old age. Society has been managing this transition from one stage to another successfully. However, this is an idealized pattern of life cycle and there are differences between the idealized and the real life, especially in today's postmodern world. There is a breakdown of traditional beliefs and practices which forces individuals to take risks and make choices. Several important life transitions–sexual initiation, partnership formation, leaving school, childbearing and entering into labour force- takes place by the time an individual reaches the second decade of his/her life. Today in an individualistic consumerist society there are impediments, which individuals have to face and negotiate in their lives. HIV/AIDS is not only one of the most salient barriers in today's world for an individual to develop oneself, as a complete human being but is also a risk in making choices. While individualism is an important aspect of today's life, the individual is part of the collectivity, society, and the product of social structures and processes, which make the social order and society. Their education, their sexual and reproductive behaviour, their readiness to take the role and responsibility of adulthood and the support they receive from the families and communities and government will determine their own futures. HIV/AIDS can also be viewed as a product of complex social structures and processes rather than the simple result of individual behaviour.

Studies on HIV/AIDS have proved this view and have been to a large extent enriched by the writings and life experiences of PLWHA. Their accounts have become politically meaningful and powerful because they have helped us understand and challenge disabling obstacles. They have been helpful because they enable us to understand more about the ways in which they work and how PLWHA have challenged social relations.

The Overview of the Response to HIV/AIDS in Darjeeling Hills

It is quite obvious that HIV/AIDS prevention is not a domain of one particular organization; it has to be conducted by more than one institution. Reaching larger number of people from all sections of the society, and carrying out effective policy work will depend on the organization's ability to network with one another. Increased collaboration could also facilitate the development of a common language to describe approaches so that practitioners can better understand one another's work. Although some networking does occur through the DACC (Darjeeling AIDS Control Center), the experience in conducting this study has shown that the networking among the different organizations could be significantly improved. Through discussions during the course of the study, it was established that the organizations involved in HIV/AIDS were not aware of others working in the same area. In spite of this, there are glimpses of efforts being made to improve on networking among the different organisations.

Being the funding agency as well as the bridge between the government and the local institutions, confronting the epidemic, DACC was under tremendous pressure to perform. However, the government has failed to understand this and it was reflected by the fact that the DACC was understaffed and for most of the executive members the work and responsibility of DACC was additional to their primary profession. As a consequence the executive members have little time to devote to the functioning of DACC. Because of the lack of staff the DACC was not able to generate commitment and sense of urgency among the people and there was a high degree of denial regarding HIV/AIDS among the community members. The disbursement of funds to other Non Governmental Organisations (NGO), working in the hills, was slow and this had become a major obstacle in responding effectively to the epidemic. The inability of the DACC to release funds on regular basis had generated immense pressure on the NGOs' effective functioning. One official from Shanker Foundation said, *"I have been coming here for the past 5 days but they do not have the cheque ready. I cannot waste my time coming here everyday as I have many other organizational works to do as well."*

There was very little involvement of the other sectors in the prevention, control, care and treatment of HIV/AIDS in the hills of Darjeeling. The tea gardens and the railways that are the major employers here are not involved in the control and prevention of HIV/AIDS. Though the railways have many programmes for HIV/AIDS across the country they were not seen to be involved here. The tea gardens do not have any programmes on HIV/AIDS for its workers. In such a situation the DACC must encourage these two big sectors to get involved. HIV/AIDS largely remains a health programme devoid of any attempt to bring about changes in the social, cultural, political and economic sphere, though there are large volumes of literature linking HIV/AIDS to the social, cultural, political and economic conditions of an individual.

In Darjeeling hills there was only targeted intervention among the high-risk groups and the community at large, which is being infected and affected by the epidemic, is being forgotten. Even now it is a taboo to talk about sex and HIV/

AIDS in the family. In such a scenario it becomes necessary for the institutions working in the sphere of HIV/AIDS to also include the community members in all the stages of planning and implementation of the programme so that the entire population is concerned about the adversity that can be brought by the epidemic.

There are many NGOs working in the field of HIV/AIDS in Darjeeling and there is a serious gap in communication among them. Though all the NGOs are linked with the DACC and Shanker Foundation, this link is very weak and merely fulfills their professional needs rather than to create stronger collaboration. The NGOs are linked with DACC for the funds and with the Shanker Foundation to fulfill the principle of GIPA (Greater Involvement of People Living With HIV/AIDS). Communication is a major training challenge to all the NGOs in the hills of Darjeeling. The communication among the NGOs at best can be said to be casual. Because of this there is not much sharing of information and data which hampers the overall future plans and programmes seriously.

All the programmes in the hills of Darjeeling are being funded by DACC. The executive members of the DACC, who decide whether or not to fund a programme, scrutinize the proposals from the NGOs. Under such circumstances the institutions are carrying out only those programmes, which have been conducted elsewhere. They do not undertake custom made programmes to suit local needs, as they fear that their proposals may be rejected. The programmes are largely high risk group centric and do not concentrate on the general public. However working with the high-risk groups is a difficult task for the NGOs as many of them are hidden and invisible. The FSW and the MSM or the bisexuals in the hills of Darjeeling cannot be identified and so working with them becomes a challenge for the NGOs. It is more of a challenging proposition for all the institutions working on issues related to HIV/AIDS as there is the absence of an enabling environment for these high-risk groups to come forward and avail the services provided by the various institutions.

In Darjeeling hills all the NGOs are concentrating only on the prevention and control aspect of the epidemic. The rural areas are left out of these prevention and control programmes to a large extent. They do not focus on care and treatment and support facilities and even the VCCTC refers the positive people to Shanker Foundation for care and support. As the intervention from the NGOs is at the level of high-risk groups the NGOs have not focused on the general public and the families of those infected by the virus. The families of the infected persons are not receiving any support from the NGOs and are going through emotional trauma. There is a greater need for intervention from the NGOs at the familial level. As many of the PLWHA have not yet disclosed their status to their families it also becomes difficult for the NGOs to approach them.

Although the number of NGOs in Darjeeling is increasing, their reach is limited only to a small segment of the society. Apart from the IRCS (Indian Red Cross Society) and the drop-in centers of MANAS Bangla and YMDS there are no independent NGOs working in Kurseong town. At the same time it can be said that a large segment of the society are unaware of the available assistance, as most of the intervention programmes have not focused on them.

Conclusion: HIV/AIDS and Human Security in Darjeeling

The aspect of human security is very important while dealing with HIV/AIDS in Darjeeling. Though the impact of HIV/AIDS in Darjeeling has not been severe enough to warrant a study of human security but it will not take much time for such a study to be a necessity. People here in Darjeeling are facing insecurity in terms of decreasing agricultural land, rapidly decreasing job opportunities and lack of proper health infrastructure. HIV/AIDS has a severe impact on food security, affecting all of its dimensions: availability, stability, access, and utilization. One of the most devastating impacts of the AIDS epidemic is the loss of human capital, as the disease robs households of adult labour and knowledge. The agricultural sector (including the tea industry) in Darjeeling will be under a particularly severe strain as the HIV/AIDS epidemic spreads, affecting and infecting more people. Given its effects on agricultural production and other livelihood strategies, HIV/AIDS will lead to a reduction of the amount of food available to individuals, households and communities. As the agricultural sector is a labour intensive sector the health of the youth is of paramount salience and their inability to work will impact the production which will directly and indirectly have an impact on the lives of thousand of individuals. This may lead to lower food intake, thus reducing both individual nutritional status and household food security.

The people of the hills are also insecure about their employment. As already mentioned earlier, many people engaged in tea industry are employed as casual labourers and some only during the plucking season they don't have a steady source of income. There is a constant threat to their wellbeing and consequently many migrate to other parts of the country and expose themselves to the risk of HIV/AIDS. Apart from the insecure nature of employment in the tea industry, Jobs in other sectors have also witnessed rapid casualization. After the formation of DGHC in 1989, the employment scenario has not improved but has in fact further deteriorated. The state government has virtually stopped recruiting people in the hills and people recruited are only on daily wage basis. This casualization of labour in Darjeeling has made the people very insecure in terms of employment and their economic condition.

Prevalence of HIV/AIDS is determined by the policies and magnitude of development. Its severity is associated with various dimensions of poverty. Poverty not only increases the chances of HIV transmission, it also increases the sufferings of PLWHA. In the wake of globalisation the present model of development promotes rapid migration from rural to urban areas. Migration from Darjeeling is also facilitating the growth of HIV/AIDS cases in the hills. Many of the female PLWHAs reported to be infected by their husbands who were working in cities.

With the growing number of HIV/AIDS cases, safe sexual practices, safe injecting practices and responsible individual behaviour is important to check the transmission of HIV. Among the people of Darjeeling responsible behaviour, with their sexual partners, including FSWs is essential to control the spread of epidemic. Ensuring individual responsible behaviour is difficult if other spheres of the society do not reflect responsibility in their intent and content. Along with the individual social responsibilities, there is need for other domains also to show responsibility towards society. State should be responsible towards the wellbeing of the people. Health system should reflect concerns towards providing universal access and affordable healthcare services. Also doctors should show commitments towards the patients. Community should reflect on the socialisation processes and societal values, and also be responsible for its members (Ritu Priya 2003).

References

ActionAid, (2006), *Positive Voices: Emerging Governance Issues on HIV and AIDS in Asia,* Shivam Sundaram, New Delhi.

Aggleton, P., (2001), "Gender and HIV/AIDS: What Do Men Have To Do With It?," *Current Sociology,* Vol 49, No 6, November, pp 23-37.

Ainsworth,M. and Teokul, W., (2000), "Breaking the Silence: Setting Realistic Priorities for AIDS Control in Less-Developed Countries," *Lancet, Vol 356,* pp 55-60.

Altman, D., (1999), "Globalisation, Political Economy and HIV/ AIDS," *Theory and Society,* Vol 28, pp 559-584.

Amathya, S., (2005), "Gender and HIV/AIDS in Nepal: Some Observations," *Indian Anthropologist, Vol 35, No1-2,* March/ Sptember, pg 139-152.

Bagla, P., (2007), "India's New HIV/AIDS Number," *EPW, Vol 42, No 30,* 28 July-3 August, pp.3080-3081.

Barnett, T., (2004), "AIDS in India: Disaster in the Making," *EPW, Vol 39,* No 39, 8-14 May p.1888-1892.

Bor, R., & Elford, J. ed, (1996), *The Family and HIV*, Cassell, New York.

Brinkhof, M. WG et al., (2008), "Early Loss of HIV Infected Patients on Potent Antiretroviral Therapy Programmes in Lower Income Countries," *Bulletine of the WHO, Vol 86,* No. 7, pp.559-567.

Burtney, E and Duffy, M ed (2004), *Young People and Sexual Health: Individual, Social and Policy Contexts,* Palgrave, New York.

Campbell, C.A., (1991), "Prostitution, AIDS, And Preventive Health Behavior," *Social Science and Medicine,* Vol. 32, No. 12 pp.1367-1378.

Carballo, M., & Siem, H., (2002), "Migration, Migration Policy and AIDS" in Haour-Knipe & Rector (ed), *Crossing Borders: Migration, Ethnicity and AIDS*, Taylor & Francis e-Library, London, www. gigapedia.org, retrived on 04.09.2007.

Catalan, J., Sherr, L., & Hedge, B., (2005), ed, "The Impact of AIDS: Psychological and Social Aspects of HIV Infection," Taylor & Francis e-Library, Amsterdam, www. gigapedia.org, retrived on 04.09.2007.

Collins,J and Rau, B., (2000), "AIDS in the Context of Development," *Social Policy and Development Programme Paper, No. 4,* December, *United Nations Research Institute for Social Development*, http://www.unrisd.org retrieved on 26.02.2006

Collins, P. Y., Unger, H. V., Armbrister, A., (2008), "Church Ladies, Good Girls, and Locas: Stigma and the Intersection of Gender, Ethnicity, Mental Illness, and Sexuality in Relation to HIV Risk," *Social Science & Medicine, Vol. 67.* No. 3, August, pp.389-397.

Craddock, S., (2007), "Market Incentives, Human Lives and AIDS Vaccines," *Social Science & Medicine,* Vol 64, No 5, March, pp.1572-1584.

Deacon, H & Boulle, A., (2007), "Commentary: Factors Affecting HIV/AIDS- Related Stigma and Discrimination by Medical Professionals," *International Journal of Epidemiology,* Vol 36, No 1, Feb, pp.185-186.

DeJong, J., (2003), *Making an Impact in HIV and AIDS: NGO Experiences of Scaling Up,* ITDG Publishing, London.

Deng, R., Li, J., Sringernyuang, L., Zhang, K., (2007), Drug Abuse, HIV/AIDS and Stigmatisation in a Dai Community in Yunnan, China," *Social Science & Medicine,* Vol 64, No 8, April.

Ford, N., and Koetsawang, S., (1991), "The Socio-Cultural Context of the Transmission of HIV in Thailand," *Social Science and Medicine,* Vol 33, No 4, pp.405-414.

Godbole, S & Mahendale S (2005), "HIV/AIDS epidemic in India: Risk Factors, Risk Behaviour & Strategies for Prevention & Control," *Indian Journal of Medical Research,* Vol. 121, April, pp.356 368.

Gupta,I., Panda,S., & Motihar, R., (2003), *HIV/AID S and Development in India, SIDA,* New Delhi.

Jacobson, N., (2007), "Dignity and Health: A review," *Social Science & Medicine,* Vol 64, No 2, January, pp.292-302.

Kalichman, S.C., (1998), *Understanding AIDS: Advances in Research and Treatment, American Psychological Association,* Washington.

Lama, D. S. & Chakraborty, C., (2007), *Cultural Diversity and Its Impact on the Hill Women of Darjeeling Hills (1835-1900),* in Victor, P. J., et al., (ed), *Discursive Hills: Studies in History, Polity and Economy,* Darjeeling, St. Joseph's College.

Mahal, A., (2004), "Economic Implications of Inertia on HIV/AIDS and Benefits of Action," *EPW*, *Vol 39,* No 10, 6 March, pp.1049-1063.

Masanjala, W,. (2007), "The poverty-HIV/AIDS nexus in Africa: A Livelihood Approach," *Social Science & Medicine, Vol 64,* No 5, March pp.1032-1041

Nelkin, D., Willis, D.P, Parris S.V. ed., (1991), *Cultural and Institutional Responses to AIDS,* Cambridge University Press, New York.

Oppong, J., (1998), "A Vulnerability Interpretation of the Geography of HIV/AIDS in Ghana 1986-1995," *The Professional Geography,* Vol 50, No 4, pp.437-448.

Oskamp, S., & Thompson, S.C. (ed), (1996), *Understanding and Preventing HIV Risk Behavior: Safer Sex and Drug Use*, Sage Publications, New Delhi.

Rhodes, F & Malotte, C.K., (1996), "HIV Risk Interventions for Active Drug Users: Experience and Prospects," in Oskamp, S & Thompson S.C. ed, *Understanding And Preventing HIV Risk Behavior: Safer Sex and Drug Use,* Sage Publication, New Delhi.

Ridge, D., et al., (2007), "Positive prevention: Contemporary issues facing HIV positive people negotiating sex in the UK," *Social Science and Medicine,* Vol 65, No 4, August, pp.755-770.

Ritu Priya, (2003), "Health Services and HIV Treatment: Complex Issues and Options," *EPW, Vol 38,* No 50, 13-19 Dec, pp.5227-5232.

Sambamoorti, U. et al., (2004), "Antenatal Care and AIDS Knowledge Among Married Women", *EPW, Vol 39,* No 7, 14-20 Feb, pp.721-727.

Samson, C., ed, (1999), *Health Studies: A Critical and Cross Cultural Reader,* Routledge, London

Santhya, K.G. and Jejebhoy, S., (2007), "Early Marriage and HIV/AIDS: Risk Factors Among Young Women in India," *EPW,* Vol 42, No 14 7-13 April pp.1291-1297.

Sarkar, S., Mookherjee, P., & Roy, A., (1991), "Descriptive Epidemiology of Intervenous Herion Users- A New Risk Group for Transmission of HIV in India," *Journal of Infection,* Vol 23, pp.201-207.

Sarkar, S., Das, N., Panda, S., Naik, T.K., Sarkar, K., Singh, B.C., Ralte, J.M., Aier, S.M., & Tripathy, S.P., (1994), "Rapid Spread of HIV Among Injecting Drug Users in the North Eastern States of India," *Bulletin on Narcotics,* Vol XLV, pp.91-105.

10

Right to Free and Compulsory Education: A Forthcoming Challenge for Sikkim

Abdul Hannan

Basic Provisions covered under the Act

The Right of Children to Free and Compulsory Education Act 2009 received the assent of the President on 26th August 2009 and thereby is a constitutional right for children in the age group of six to fourteen years. Now, it is the obligation and compulsion of State and Central Governments to provide every child of the age of six to fourteen years, male and female, with free and compulsory education in a neighbourhood school till completion of elementary education under section (3) and clause (1). The Act also talks of out-of-school children, wherein a child above six years of age who has not been admitted in any school, or though admitted, could not complete his or her elementary education. Such children shall be admitted into a class appropriate to his or her age under section (4) and shall have the right to receive special training to become at par with others. The Act also ensures a screening procedure for admission and bars the acceptance of any 'capitation fee', that is, any kind of donation or contribution or payment other than the fee notified by the school. Any such offence is punishable with fine extending to twenty-five thousand rupees for the first contravention, and fifty thousand rupees for each subsequent contravention.

The basic facts covered under the Right to Free and Compulsory Education Act may be summarized as follows:

Quality Education

It includes stipulations about a specified Teacher-Pupil Ratio, minimum qualification of teachers, minimum number of working days and instructional hours in an academic year, minimum number of teaching hours in a weak, achievement of 100 per cent Trained Teachers, establishment of Library in Primary and Upper Primary schools, regular provision of teaching-learning equipments for each class, comprehensive and continuous evaluation of the child's understanding of knowledge, and the provision of sports materials and playground access for the all round development of children at the elementary level.

Social Responsibility or Inclusion

Private schools shall have to reserve twenty-five per cent of their seats for children belonging to *disadvantaged groups* and *weaker sections,* living in the neighbourhood of the school.

Financial Responsibility

It will be the concurrent responsibility of both the Centre and the States to provide funds with the Central Government preparing the estimates of capital and recurring expenditure for the implementation of the provisions of the Act. The Central Government shall also provide to the State Governments as grants-in-aid of revenue, principal, as determined from time to time in consultation with the latter.

De-bureaucratisation of School Management

The School Management Committee shall consist of the elected representatives of the local bodies, parents or guardians of children admitted in schools and teachers. The manner of participation is a) Three-fourth of the members of the committees shall be parents or guardians of children with proportionate reservation for the parents or guardians of children belonging to disadvantaged groups and weaker sections; and b) Fifty per cent members shall be women. The main objective of the School Management Committee shall be to prepare and recommend the School Development Plan, which shall be the basis for the plans and grants for the schools.

Appropriate Physical Infrastructure

Physical infrastructure refers to all-weather buildings, at least one classroom per teacher, a store-cum Head Teacher's room in Primary and Upper Primary Schools, road connectivity and barrier-free access, separate toilet for boys and girls, safe drinking water facility, kitchen for mid-day meal preparation, a playground and the boundary wall of the school.

Duties of Appropriate Government, Local Authority and Parents

- Every child of the age of 6-14 years shall have the right to free and compulsory education in neighbourhood institutions/schools.
- Appropriate Government or Local Authority shall establish schools in the neighbourhood within a period of three years from the commencement of the Act.
- Appropriate Government or Local Authority may make necessary arrangement for providing free pre-school education for the children above the age of 3-6 years.

- Every parent or guardian shall have to admit or work for the admission of his or her child/ward in the neighbourhood school.

Norms and Standards for setting up of Schools

- No school shall be established unless it fulfills the norms and standards as specified in the schedule of the Act.
- Where a school established before the commencement of the Act and doesn't fulfill the norms and standards, such institutions have to meet all norms and standards within a period of three years.
- Central Government, by notification, may amend the Schedule of norms and standards.
- School Management Committee shall consist of elected representatives of the local authority, parents or guardians, and teachers in manner as follows: a) Three-fourth of members shall be guardians or parents; b) further proportionate representation of the guardians or parents belonging to disadvantaged groups or weaker sections; c) Fifty percent members of women.
- Preparation of School Development Plan.

Recruitment Policy, Duties and Responsibilities of Teachers

- Any person possessing minimum qualification, laid down by academic authority, authorised by the Central Government by notification, is eligible for appointment of teacher.
- In the case of non-availability of trained teachers in sufficient numbers, the provision of relaxation of minimum qualification for appointment of teachers subjected for a period of not more than five years and

teachers shall acquire such minimum qualification within a period of five years.

- Appropriate Government or Local Authority shall ensure Teacher-Pupil Ratio within six months from the date of commencement of the Act.
- Vacancy of Teachers in a School shall not exceed ten percent of the total sanctioned strength.
- Non-deployment of teachers for any non-educational purpose except Census Operations, Disaster Relief Management and Elections (Local, State, and Parliament).
- Non-engagement of teachers in private tuition or private teaching activity.

Forthcoming Challenges for the State of Sikkim

The Right of Children to Free and Compulsory Education Act, 2009 has made a provision to provide for free and compulsory education to all children of the age of 6-14 years and it considers the Pre-Primary, Primary and Upper Primary levels of education and all these three categories are considered under the elementary level of education. In Sikkim, four types of management - State Government, Central Government, Government Aided and Private schools - provide elementary education. During the 2008-09 session, there were 142,018 children (students) enrolled in elementary education, out of which 29,415 (which is 20.71 %) were in the Pre-Primary, 81,366 (57.29 %) in the Primary and 31,237 (22.00 %) in the Upper Primary levels. As much as 72.30 % of the total numbers of students enrolled in elementary education were under the management of the State Government, followed by Private Schools (25.11%), Government Aided (1.68 %) and the Central Government (0.90 %).

Table1
Levels of education, enrolment status and management (2008-09)

Levels of Education	*Number of Students Enrolled*				
	State Govt.	*Central Govt.*	*Govt. Aided*	*Private Schools*	*Sikkim*
Pre Primary	14758	210	390	14057	29415 (20.71)
Primary	62465	476	1098	17327	81366 (57.29)
Upper Primary	25463	593	904	4277	31237 (22.00)
Elementary Level	102686 (72.30)	1279 (0.90)	2392 (1.68)	35661 (25.11)	142018 (100.00)

Note: Figures in the parentheses are % share of elementary education (last column); and % share of management of elementary education (last row).

Source: Computed from Unpublished Data collected from Human Resource Development Department, Government of Sikkim.

In the State as a whole, there were 575 elementary education schools run by the State Government and 375 managed by Private institutions (Table-2). The average number of students per School was 179 and 97 in Government and Private respectively. In fact, within the elementary level there were variations of average number of students per school as well. The average number of students per school in Pre-Primary (102) was far below the average number of students of the total elementary education level (179). The average number of students per school in Upper Primary level (174) was just below the average of elementary education. However, the average of Primary level (219) was very much above the average of elementary education. On the other hand, in Private Schools this picture was totally different (187 in the Pre-Primary level, 89 in the Primary level, and 43 in the Upper Primary level). The highest average size of students per school was found in the Primary level in Government schools while in Private schools it was found in the Pre-Primary level. It was also observed that as the level of education rose, the average number of students per school fell sharply in the case of Private Schools.

Table 2
Number of schools and students and average size of the school (2008-09)

Levels of Education	*State Government*			*Private Schools*		
	No of Schools	*No of Students*	*Average Number of Student Per School*	*No of Schools*	*No of Students*	*Average Number of Student Per School*
Pre Primary	144	14758	102	75	14057	187
Primary	285	62465	219	194	17327	89
Upper Primary	146	25463	174	98	4277	43
Elementary Level	575	102686	179	367	35661	97

Note: Average number of Student per School is calculated as number of Students divided by number of Schools.

Having discussed the basic facts of elementary education in the State, the forthcoming challenges can be grouped into four categories, namely, human resources, physical infrastructure, policy inclusion of private institutions, and additional financial requirement.

Human Resources

Trained Manpower in Sikkim

As per the definition of NCERT, "A trained teacher is one who has successfully undergone a course of training through a recognized teacher training institution, i.e. B.Ed./BT, JBT, ETTE/NIT/SV/CT/JEE/LT/BM/PET etc. A deemed trained teacher is one who has been awarded a certificate or exempted from training by the Department of Education on the basis of experience/age etc" (www.ncert.nic.in). As per Annual Report 2005-06, in Sikkim, 72.43 % of Teachers are trained at Primary level and 42.48 % at Upper Primary level. Almost 30 % in Primary and 60 % in Upper Primary level are untrained. As per the provision of the Act under section (23) clause (2) it is again compulsory for the State to achieve 100% Trained Manpower within a period of five years from the commencement of the Act.

Dropout Rates

The concept of dropout reflects the internal inefficiency of the education system and roughly indicates the wastage involved in the school education. Various factors coexist and interact including economic compulsions and social pressures in the process of dropping out of school-going children (Sikkim Development Report: 2008:79). As per the definition of NCERT, "A dropout is a pupil who leaves school before the completion of a school stage or leaving at some intermediate or non-terminal point of a given level of education (School Stage). This term 'dropout' has been used in two senses. It may mean either: (i) one who has discontinued education before completing the last level of education for which he/ she was enrolled or (ii) one who has discontinued education before attaining a specific level" (www.ncert.nic.in). In Sikkim, the dropout rate is 22.02 per cent at elementary level and 12.59 per cent in secondary level. In other words, if 100 students were enrolled in Class-I in the year 2000, only 78 students have reached Class-VIII in 2007 and the other 22 students have dropped out from the elementary education. Under the Act every child of the age of six to fourteen shall have the right to free and compulsory education in a neighbourhood school till the completion of elementary education under section (3) and clause (1), and further it says that no child shall be liable to pay any kind of fee or charges or expenses which may prevent him or her from pursuing and completing elementary education. Therefore, the State Government has to have an adequate number of schools and other related facilities, and also a pragmatic approach to the eradication of poverty, which will ultimately lead to the reduction of the dropout rate to zero per cent in due course of time.

Table 3
Dropout rate in Sikkim

Levels of Education	*2007*			*2005*		
	Boys	*Girls*	*Total*	*Boys*	*Girls*	*Total*
Elementary Level (I to VIII)	21.63	22.42	22.02	23.02	21.70	22.39
Secondary Level (IX to X)	12.86	12.36	12.59	12.00	16.66	11.35
Total	24.78	26.11	25.42	27.13	25.39	26.29

Source: Compiled from Draft Annual Plan 2009-10, Part-II, December 2008 & Sikkim Development Report (2008).

Teacher-Pupil Ratio

The State of Sikkim has one of the lowest Teacher-Pupil Ratios in the country, which means that the State has high number of teachers per student or pupil. The teacher-pupil ratio at Primary level in Sikkim is 1: 18 as against 1: 34 in India (2004-05) (http://www.sikkim.gov.in). The district-wise distribution of the ratio ranges from 1: 17 to 1: 21, which again is low in comparison to the National average. But as per the Schedule of the Act, if a school has above 150 students it should also have a Head Teacher. The average number of students per school is 219 (Table-2) in Government Primary Schools in the State, which reveals that all the Government Primary Schools should essentially have Head Teachers as per the Schedule under section (25) and clause (1) and this condition is to be met within six months from the date of commencement of the Act.

Table 4
Teacher-pupil ratio in primary level (I to V)

Norms & Standards as Per Schedule of the Act for Primary Level		*Existing Scenario of Primary Level in Sikkim (2005)*	
No. of Students	*No. of Teachers Required*	*Districts*	*Teacher-Pupil Ratio*
Upto 60	2	East	1:17
61-90	3	West	1:21
91-120	4	North	1:18
121-200	5	South	1:18
Above 150	5+1 HT*	Sikkim	1:18
Above 200	1: 40 + HT*	All India	1:34

*HM: Head Teacher
Source: Compiled from the Act and Annual Report 2005-06, HRDD, Government of Sikkim.

Similarly, at the Upper Primary level the Teacher-Pupil Ratio is low in Sikkim and it ranges from 1: 22 to 1: 24 in different districts. As per guidelines of the Schedule of the Act, the ratio in Upper Primary should be 1: 35 and the State of Sikkim fulfills this. Further, the Act also mentions the requirement of various subject specializations of the Teachers in a school and it is mandatory to have a minimum of three teachers at Upper Primary level from Science & Mathematics, Social Studies, and Languages (Table-5). Besides these, if a school has more than 100 students, it specifies to have a Head Teacher and Part-Time Instructors for Art Education, Health Education, Physical Education, and Work Education for the all round development of children. In Sikkim, the average number of students per Upper Primary Government School is 174 (Table-2), which clearly implies that all the Upper Primary Schools run by the State Government in the State should essentially have Head Teacher and Part-Time

Instructors. All these conditions are to be met within six months from the date of commencement of the Act and State shall ensure the Teacher-Pupil Ratio as per the Schedule under section (25) and clause (1).

Table 5

Teacher-pupil ratio in upper primary level (VI to VIII)

Teacher-Pupil Ratio as per Schedule of the Act for Upper Primary Level	*Existing Scenario of Upper Primary Level in Sikkim (2005)*	
	Districts	*Teacher-Pupil Ratio*
A) At least 1 Teacher/Per Class consisting 1 Teacher each in		
(i) Science & Mathematics	East	1:23
(ii) Social Studies	West	1:24
(iii) Languages.	North	1:16
B) At least 1 Teacher/35 Students.		
C) Above 100 Students		
(i) 1 full-time Head Teacher,	South	1:22
(ii) Part-time instructors for	Sikkim	1:22
a) Art Education;		
b) Health & Physical Education;		
c) Work Education		

Source: Compiled from the Act and Annual Report 2005-06, HRDD, Government of Sikkim.

There are various responsibilities given to the Teachers as well as State Government that are mentioned in the Schedule of the Act. As per the section (26), the State Government shall have to ensure that the number of vacancies of teachers shall not exceed the total sanctioned strength and that the sanctioning of vacancies should be in accordance to the Schedule of the Act. Under section (27) of the Act, teachers cannot be deployed for any non-educational purposes except Census operations, disaster relief and the Local, State and Parliamentary elections. Moreover, under section (28) of the Act, no teacher can be engaged himself or herself in private tuition or private teaching activity. The norms of working days in schools is 200 days for Primary level and 220 days for Upper Primary level in an academic year and 800 Instructional hours for Primary level and 100 Instructional hours for Upper Primary level. The teaching-cum-preparation is prescribed as 45 hours in a week for the Teachers in both the Primary and Upper Primary level. State Government shall have to ensure the implementation of the norms and standards set under the Act.

Table 6
Norms for minimum number of working days/ instructional hours and working hours

Item	*Working Days / Instructional Hours and Working Hours*
A) Minimum number of Working Days/Instructional Hours in an Academic Year.	• 200 Working Days for Primary Level. • 220 Working Days for Upper Primary Level. • 800 Instructional Hours for Primary Level • 1000 Instructional Hours for Upper Primary Level
B) Working Hours for Teachers in a week.	• 45 Teaching including preparation Hours.

Source: The Right of Children to Free and Compulsory Education Act, 2009

Physical Infrastructure

As mentioned earlier in Table-2 that there are 285 and 146 number of Primary and Upper Primary Schools run by the State Government in Sikkim. Out of which 22.33 % of Primary and 10.08 % of Upper Primary Schools still have *Kutchcha* buildings. This leads us to conclude that that there are 64 schools in Primary level and 15 schools in Upper

Primary level that have *Kutchcha* buildings, while the Act also elaborately explains about other physical infrastructures such as one classroom per teacher and a store-cum Head Teacher's room, road connectivity and barrier-free access, separate toilet for boys and girls, safe drinking water facility, kitchen for mid-day meal preparation, playground and boundary wall of the school. As per section (19) and clause (2), in the case of a school established before the commencement of this Act that does not fulfill the norms and standards specified in the Schedule, the State Government shall have to take steps to fulfill such norms and standards within a period of three years from the date of commencement of this Act.

Table 7
Norms for buildings and other physical infrastructures

Building and other Physical Infrastructures as per Schedule of the Act in Primary and Upper Primary Level	*Existing Scenario of Buildings in Sikkim (2005)*	
	Levels of Education	*Kutchcha Buildings (%)*
All weather building consisting of:		
• At least 1 Class-room for every Teacher and an office-cum-store-cum-Head Teacher's room. • Barrier-free access. • Separate toilets for boys and girls.	*Primary*	22.33
• Safe & adequate drinking water facility. • A kitchen for mid-day meal preparation. • Playground. • Boundary wall or fencing for securing the school.	*Upper Primary*	10.08

Source: Compiled from the Act and Annual Report 2005-06, HRDD, Government of Sikkim.

Teaching-Learning aids and equipments, and sports materials of various kinds are to be provided to schools in Primary and Upper Primary levels. It is also compulsory to have a Library in Primary and Upper Primary Schools. Newspapers, magazine and books on all subjects and storybooks should be made available. The State Government has to ensure and provide from time to time such basic facilities and establish Libraries in all the schools in the State.

Table 8
Miscellaneous Norms and Standards as per Schedule of the Act

- Teaching-learning equipments and materials shall be provided to each class.
- Play material, games and sports equipment shall be provided to each class.
- Shall have the Library in each school providing newspaper, magazine and books on all subjects including storybooks.

Source: The Right of Children to Free and Compulsory Education Act, 2009.

Access to Education

As per the definition of the NCERT, "The distance of a school from a rural habitation is the convenient walking distance between the central point of the rural habitation and the school. A rural habitation is a distinct cluster of houses existing in a compact and contiguous manner; with a local name; and its population should not be less than 25 in plain areas and not less than 10 in hilly/desert/sparsely populated areas. In case there exists more than one such cluster of houses in a village, they will not be treated as separate habitations unless the convenient walking distance between them is more than 200 metres." Educational facilities within rural habitations are defined by walking distance and it is measured for Primary Schools within 1 km of walking distance and for Upper Primary Schools within 3 km of walking distance. As per Annual Report 2005-06 (HRDD, GoS), 87.28 % of rural habitations have Primary Schools within 1 km of distance and 83.26 % of rural habitations have Upper Primary Schools within 3 km of distance. In Sikkim, there are 901 habitations out of which 786 habitations have access to Primary Schools and 750 habitations have access to Upper Primary Schools. On the contrary, 115 habitations do not have access to Primary schools and 151 habitations do not have Upper Primary schools. Under section (6) of the Act, it is mandatory to establish schools within the limits of the neighborhood within a period of three years from the commencement of this Act.

Policy Inclusion of Private Institutions

The State has a high proportion of private institutions at the elementary level of education. There are 618 Government Schools and 367 Private Schools providing elementary level education, that is, 63 % and 37 % respectively, in the State. The composition of Government and Private School is 66 % and 35 % at Pre-Primary level, 59 % and 41 % at Primary level, and 60 % and 40 % at Upper Primary level. The dimension of regional distribution at the district level in terms of the public-private participation is clearly evident from Table-9. There are 129 private schools in the East district followed by 109 in the West, 108 in the South and 62 in the North. The highest concentration of private schools being in the East district is somehow related to the location of the State Capital i.e. Gangtok.

Though over the years, private schools have minimized the burden of the State Government by sharing the responsibility of elementary education, but from now onwards it is mandatory for the State to make provisions for the twenty-five percent enrolment of children belonging to weaker sections and disadvantaged group in the neighbourhood in these private schools, under the section 12

(1) clause (c). In addition, it also mentions that wherever a school imparts pre-school education, this clause shall apply for admission to such pre-school education as well. The school shall be reimbursed expenditure incurred to the extent of per-child-expenditure incurred by the State or actual amount charged from the child, whichever is less, under section 12 (2). The responsibility of the State Government is to monitor and prepare a guideline for private schools and it has to complete this process within a period of three years from the date of commencement of the Act. If any schools fail to adhere to the guidelines, the State shall withdraw recognition granted to such schools.

Table 9
Ratio of public and private schools in elementary education (2008-09)

Levels of Education	*Number of Schools and their % Share*	*East Govt. Private*	*West Govt. Private*	*North Govt. Private*	*South Govt. Private*	*Sikkim Govt. Private*
Pre Primary	No of Schools % Share	27:24 (53: 47)	56:19 (75: 25)	15:60 (71: 29)	46:26 (64: 36)	144:75 (66: 34)
Primary	No of Schools % Share	100:74 (57: 43)	79:55 (59: 41)	24:33 (42: 58)	82:56 (59: 41)	285:194 (59: 41)
Upper Primary	No of Schools % Share	47:31 (60: 40)	34:35 (49: 51)	17:23 (43: 57)	48:26 (65: 35)	146:98 (60: 40)
SSA*		10	13	13	7	43
Elementary	No of Schools % Share	184:129 (59: 41)	182:109 (63: 37)	69:62 (53:47)	183:108 (63: 37)	618:367 (63: 37)

*Schools established under Sarva Shiksha Abhiyan.

Source: Computed from Unpublished Data collected from Human Resource Development Department, Government of Sikkim.

To fulfill this mandate, the *Gross Enrolment-cum-Social Inclusion Ratio* gives a clearer picture of the situation in the State of Sikkim. During 2008-09, there were 35,661 children enrolled in the private schools and as per the provision of the Act, 8,915 of these seats (which is 25 %) should be allotted to weaker sections and disadvantaged group in the neighbourhood under the above-mentioned clause (Table-10). Similarly, 3514 such children in the Pre-Primary level, 4332 in the Primary level, and 1069 in the Upper Primary level are to be admitted, as per provisions. At the regional level, the number of children should be 4256 in East district, 2101 in West, 542 in North and 2015 in South.

Table 10
Gross enrolment-cum-social inclusion (SI) ratio in private institutions (2008-09)

Levels of Education	*Number of Children*				
	*East Gross: SI**	*West Gross: SI**	*North Gross: SI**	*South Gross: SI**	*Sikkim Gross: SI**
Pre Primary	6292:1573	3401: 850	1008: 252	3356: 839	14057: 3514
Primary	8254: 2063	4182: 1046	957: 239	3934: 983	17327: 4332
Upper Primary	2480: 620	821: 205	205: 51	771: 193	4277: 1069
Elementary	17026: 4256	8404: 2101	2170: 542	8061: 2015	35661: 8915

Note: Gross Enrolment refers to total number of admitted children, and Social Inclusion (SI) refers to twenty-five percent of Gross Enrolment.

Source: Computed from Unpublished Data collected from Human Resource Development Department, Government of Sikkim.

Additional Financial Requirement and Budgetary Provision

The Table-11 represents a picture of budgetary allocation and expenditure over different plan periods. It is easily observed that the highest share of the budget is allocated to elementary education in all Plans. Its share is 47.55 %, 36.60 %, and 40.83 % in 9th, 10th and 11th Plan periods respectively. Once the Right of Children to Free and Compulsory Education Act, 2009 is implemented, State Government will require additional financial resources to meet the expenditure for fulfilling the norms and standards set under the Act. Under

Table 11

Budgetary allocation and expenditure in education

Sl. No.	Level of Education	9th Plan (Rs. in Lakhs)		10th Plan (Rs. in Lakhs)		11th Plan (Rs. in Lakhs)	% Outlay during different Plan Periods		
		Approved Outlay	Actual Exp.	Revised Outlay	Actual Exp.	Approved Outlay	9th Plan	10th Plan	11th Plan
1	Elementary Education	9509.00	11214.10	14859.70	14918.27	26750.00	47.55	36.60	40.83
2	Secondary Education	8070.00	5824.48	10647.70	11659.33	25650.00	40.35	26.23	39.15
3	Univ & Higher Education	1622.00	943.51	2472.80	2216.28	6200.00	8.11	6.09	9.46
4	Adult Education	43.00	11.01	6.20	1.20	100.00	0.22	0.02	0.15
5	Language Development	90.00	20.87	35.70	42.38	100.00	0.45	0.09	0.15
6	Direction & Admn.	Nil	350.88	964.00	1050.74	1600.00		2.37	2.44
7	Scholarships	350.00	292.09	319.50	303.70	600.00	1.75	0.79	0.92
	General Education (Sl No. 1 to 7)	**19684.00**	**18656.94**	**29305.60**	**30191.90**	**61000.00**	**98.42**	**72.18**	**93.10**
8	Technical Education	316.00	161.89	11293.50	5597.22	4523.50	1.58	27.82	6.90
	Total (Sl No. 1 to 8)	20000.00	18818.83	40599.10	35789.12	65523.50	100.00	100.00	100.00

Source: Compiled from 9th Plan (Volume-I) and 11th Five Year Plan (Part-III), and Unpublished Data Collected from Planning and Monitoring Division, HRDD, Government of Sikkim.

section (7) and clause (1), the Central Government and State Governments shall have concurrent responsibility for providing funds for implementing and carrying out the provisions of the Act. The Central Government shall prepare the estimates of capital and recurring expenditure for the implementation of the provisions made compulsory under the Act and shall provide to the State Governments as grants-in-aid of revenues as determined from time to time in consultation with the State Government.

The State Government shall also have to constitute, a ***State Advisory Council,*** by notification, consisting of members not exceeding fifteen from amongst the persons having knowledge and practical experience in the field of elementary education and child development under the section (34) and clause (1). The functions of the State Advisory Council shall be to advise the State Government on implementation of the provisions of the Act in an effective manner under section (34) and clause (2).

Notes

1. The term *Students* and *Children* are synonymous in this paper.
2. Average number of Student per School is calculated as number of Students divided by number of Schools.
3. Social Inclusion refers to "child belonging to disadvantaged group" covers a child belonging to the Scheduled Caste, the Scheduled Tribe, the socially and educationally backward class or such other group having disadvantage owing to social, cultural, economic, geographical, linguistic, gender or such other factor; and "child belonging to weaker section" covers a child belonging to such parent or guardian whose annual income is lower than the minimum limit specified by the appropriate Government by notification.
4. Gross Enrolment-cum-Social Inclusion (SI) Ratio: Gross Enrolment refers to total number of admitted children, and Social Inclusion (SI) refers to twenty-five percent of Gross Enrolment.

References

Annual Reports (2005-06), Human Resource and Development Department, Government of Sikkim.

Arun C. Mehta, Indicators of Educational Development with focus on Elementary Education: Concepts and Definitions, http://www.educationforallinindia.com/New%20Modules/module%20on%20indicators%20of%20educational%20development.pdf accessed on 05/11/2009.

Draft Annual Plan 2009-10 Part-II, December 2008, Development Planning, Economic Reforms & North-Eastern Council Affairs Department, Government of Sikkim.

Sikkim Development Report, Planning Commission, Government of India, published by Academic Foundation, New Delhi, 2008.

The Right of Children to Free and Compulsory Education Act, 2009, Government of India, ***http://www.education.nic.in*** accessed on 15th October 2009.

9th Plan (Volume-I) and 11th Five Year Plan (Part-III), Development Planning, Economic Reforms & North-Eastern Council Affairs Department, Government of Sikkim.

http://www.ncert.nic.in/html/pdf/educationalsurvey/Manual_on_Statistics_and_Indicators_of_School_Education/Concepts_and_Definitions__Alphabatical.pdf accessed on 25th October 2009.

http://www.sikkim.gov.in accessed on 26th October 2009.

11

The Dynamics of Women, Work and Social Status: A Case Study of the Bhutias of Sikkim

Swati Akshay Sachdeva

Introduction

It cannot be denied that women who compose almost half of the total population, still remain the most neglected lot. Women constitute the only 'group' that is treated unequally as a whole, in spite of their comprising the largest group in the world.

There is a commonplace observation that mainstream social and political thought has commonly accepted and confirmed women's subordinate position in social and political life. Describing them as the "weaker sex" is used to legitimize this view. However, with more and more women joining various occupations this view is changing. The whole pattern of male-female roles is undergoing a considerable change. The change in social conditions and pattern of living has been responsible for women going out for jobs and this is responsible for the shift towards their 'outdoorness'.

According to another viewpoint, women did enjoy certain privileges and were dominant at some stage of evolution of society as seen in matriarchal society of Khasis in Meghalaya. But women's position has never been consistent, in some societies they occupied a high social position and in other societies at the same time they had a low social position.

The position of women in society can be explained by taking into consideration various socio-economic and socio-cultural dimensions. According to Marx (1963), the position of women will vary from society to society and time to time according to prevailing economic and political relationship. Marx placed women in mode of production of society and interprets them as forming a class that is subordinate to and oppressed by the superior class formed only by men under certain historical circumstances. Possibility of women's exploitation historically, can be explained by the argument that nations and women are not spared in their unguarded hours, particularly when a more adventurous warrior is around (Marx 1963). While developing a view on women's position, Engels (1948) argues that property management, patriarchal system and the government are all related. According to Engles, the capitalist production process is responsible for the subordinate position of women. Engels' rationale is that, human societies in all times were dependent equally on men and women in manners of production and procreation. However, it is difficult to say that whether prior to capitalist production, men and women occupied equal position.

According to anthropological point of view, the subordinate position of women is due to the process of social evolution. As society came into existence, interaction started taking place between groups, and because of interaction a material exchange of goods became possible. In this process, the women for the material exchange value attached to them, acted as a source of facilitation of social interaction between groups (Strauss 1953). Niranjna (2000) mentioned, "This material exchange takes place not merely in terms of women as a human body meant for sexual pleasure and carrying on the reproductive activities. Rather it acts as a medium of cultural patterns and practices, ancestral rights and social structural aspects of different groups between whom the exchange takes place". Hence, women were compared to physical objects, a premise that further undermined their position.

Some aspects of culture and social structure of society also helps to determine the image of women. Nature endows human beings with an unalterable biological makeup by classifying them into male and female. However, the definition of feminity and masculinity is bestowed upon humans by culture and society. In some cultures feminity is associated with a number of socio-economic, political roles, while in some other cultures feminity is almost entirely associated with functions of a domestic nature. There are perceived differences between men and women based on the kinds of relationship possible, the degree of female sexual purity and norms regarding the division of labor. The role of women is multidimensional involving spiritual, sexual, intellectual and managerial aspects but still a lot of emphasis is placed on female sexual virtue, as in most societies the sexual conduct of a women reflects on the honor of entire social conduct.

Many cultures perceive women as basically emotional and thus inferior to men. The inability of women to cope with decision-making has been used as a justification for paternalism and for the treatment of women as minors throughout their lives. It is assumed that women's intellectual aptitude is inferior to men, so they are suitable for only stereotyped jobs. Another dimension of cultural, which gives women a subordinate position in society, is power. Power, in all societies has been regarded as a male prerogative.

Irrespective of other elements in their role definition, women have been socialized to consider themselves, unsuitable for leadership or have been taught that it is not feminine to desire power (Rathschild 1974).

Religion is the source of many laws and customs in society and because of its association with rituals it marks important occasions in the life of the individual. The spiritual image of the women has many components. In religious systems with a tradition of world denial and preoccupation with salvation, such as pre- Islam Buddhism and Hinduism of Bangladesh, the Buddhism of Japan, and the Medieval Christianity of Poland and France the spiritual worth of women is depended on their faculty for achieving personal redemption (Giele and Smock1977). In India, from the point of view of religion the husband is equated to God, and as a result, women here are further subdued. Some cultures emphasize the difference between male and female without placing sexes in hierarchical order, others clearly mark male superiority and male inferiority and the third view male and female as relatively equal. In hierarchical relationships, women were given a domestic role and were under the supervision of men. The modified hierarchical system of relationship was flexible and there was smaller degree of sexual segregation (Giele and Smock 1977). However, relative equality remains a theoretical possibility.

Some major structural variables that influence the position of women in the society - like the effectiveness of the family as an institutions and -the manner in which female and male roles relate to each other, the access of girls to education on equal terms, the type of economic system and the nature of the economic opportunities it offers to women needs to be discussed because family of its centrality in social organization influence women's position. Woman always subordinates her needs to the interests of the family because she is an integral part of family system and this limits her ability to act independently. However, equal educational opportunities when given to women can have both macroscopic and microscopic implications, because it transforms the attitude of women, by endowing them with greater skills and resources. The greater the personal resource a woman brings into a marital relationship in the form of her education, employment experience and a range of social contacts – the closer her resource base approximates that of her husband – the greater is the chance that she will be able to influence family decision making (Rodman 1972).

Women empowerment depends on economic participation, which depends on her ability to accept the employment with her domestic responsibilities. Because of which, on one hand women from upper class backgrounds have more protection and prestige, whereas families of humbler origin cannot afford the luxury of excluding women from the labor force. This reveals that there are cultural, social and economical constraints of the society on the life of women.

Sikkim, situated in the Eastern Himalayas is a meeting ground of several cultures manifest in the distinctive lifestyle of its several tribal communities. According to the 2001 census, Sikkim has a population of 5, 40493. Ethnically, the state comprises of three groups mainly, the Bhutias, the Lepcha and the Nepalese.

Bhutias: A Background

The derivation of Bhutia is from Bhote or Bhot, which itself is derived from Bod or Tibet. In a Sanskrit manuscript of seventh century, Tibet is called Bhote. People of Tibetan stock who migrated to the southern face of the Himalayas are known as Bhot or simply Bhutia. Communities of such people are found in a number of areas from as far west as Uttar Pradesh to Sikkim, even though their ethnic origin may differ (Lall 1981).

The name of the Tibet is of Tatar origin in the records of the Tarter Liaos in the eleventh century the name is written *T'u-pote's* in which the latter syllable represents Bod. . The word first occurs about 950 AD in the works of an Arab writer Istakhari, who called the country *Tobbat* that is the Arabic rendering of the word *Tho-bhyodh*. The name suggests that Bhotia called the great plateau with it's a plain boarding the frontiers of China, Mongolia and Kashmir as *Tho-bhyodh* got its name Tibet in European accent.

The term Bhutia is itself confusing because it is used for an ethnic group of Sikkim and Darjeeling, as well as for the inhabitants of Bhutan. But the Bhutias of the UP hills have absolutely no connection with the communities in the Western Himalayas. They form an integrated part of local Hindu culture with a possible admixture of the Tibetan ethnic and culture elements due to long contact with them. (Lall, 1981).

Origin and Migration of the Bhutias of Sikkim

The name Bhutia, sometimes spelt and pronounced as Bhotia is an exclusive ethnic community different from which both Hindu and Tibetans. It is a generic term used to designate several socially unrelated group Indo-Tibetan borderlands. However, diverse the Bhutia may be they profess Buddhism in one form or another and speak a Language broadly intelligible to each other.

It has evidently been accepted that people from "*Kham*" province of eastern Tibet first started to come down and to settle in Sikkim from about tenth and eleventh centuries, due to various socio-political, economic and religious reasons (Bhattacharya 1994). Shortly, Sikkim became the colony of the Tibetan Settlers. By the middle of the 17th century, three venerable Lamas of Tibet came to Sikkim and established the Namgyal Dynasty in 1642 just as the prototype of the Lamaist Tibet.

The Bhutias of Sikkim as history says are of Tibetan origin, having same cultural pattern, language and scripts as the Tibetans. In the Tibetan dialect, Tibet is called Bod, which means land of snow and the people are called *Bod-pa*.

Tibet in Sanskrit literature is regarded as Vota Desh and the people of Tibet called Bhoteas or Bhutias. It is therefore assumed that Vota or Bod can be the origin of the term Bhutia.

The ethnic origin of Sikkim Bhutias is very complex and includes Drukpas, people from Drukul/Bhutan (Thunder, Dragon). The Chumbipas, people from Chumbi Valley (Southern Tibet) the Dhopthapas, inhabitants of Dhoptha, a place in South Tibet, the Tromopas or *Do-mu-pas*, inhabitants of *Du-mu,* the Lachenpa, the people of Lachen Valley in Sikkim, and the Lachungpas, the people of the Lachung Valley in North Sikkim (Singh 1993).

The Tibetan who came down to Sikkim from upper region of Tibet were devout Buddhist and brought with them Tibetan Mahayana Buddhism, known as 'Vajrayana'. The result is that the Bhutias, who came to Sikkim from Tibet, developed a social and cultural pattern different from Tibetans, so that they are classified as Sikkimese Bhutia. The Sikkimese Bhutia is one of the many tribes belonging to the corpus Bhotia, they call themselves Sikkimese or Denzongpa and speak Sikkimese which is a dialect of Tibetan.

Tibetan settlers came to Sikkim from neighboring valleys of Chumbi and Ha, and regions even beyond this from the thirteenth century. Their descendants call them Lhopo (*lho pa,* people of Denjong or Demojong (*'bras mo ljongs*, the country of rice). Since the term Bhutia can be used for any person of Tibetan origin and because they are present in a minority in Sikkim therefore the term 'Sikkimese' cannot be used for them. Hence they are called Lhopo (Balikci 2002). However in the present paper the word 'Bhutias' have been used everywhere.

To study the dynamics of women, work, and social status of Bhutia women in Sikkim, a field research was undertaken in all the four districts of Sikkim i.e. East, West, North and South. From each of the four districts, villages and areas with maximum Bhutia population was considered as an ideal choice as it provided a population with somewhat homogenous cultural setting. The level of development was also taken into consideration as it provided variation in the socio-economic status of women. A number of 50 samples (respondents) were taken randomly from the study area in each district.

From North district, (headquarter at Mangan) Tingchim was selected for the purpose of study. From the East district (headquarter at Gangtok and which also the state capital) three areas were selected i.e. Navey, Uppper Tathangchen and Tathangchen. From the South district (headquarter at Namchi), Barfung and Ralong were selected as the field. From West district (headquarter at Gyalshing), Singyang and Sindrang were taken as study areas and some samples were also collected from Gyalshing.

A Likert-type scale was used for measuring the continuous variables and this divided the respondents in the relevant cases into categories marked high, medium and low on the basis of the score obtained by them. (Kothari 2004).

Women in Work and Decision-making

The women of Sikkim are unique in many ways. "Like everything else in Sikkim they also truly manifest the influence of the sylvan serenity of the august hills of the Himalayas and the deep pervasive influence of the quintessence of Buddhism. These influences have made their lives free from many discriminatory inequalities from which the women suffer elsewhere" (Bhattachrya 1994).

The participation of women in work depends upon their education, socio – cultural values attached with work and availability of work.

The work structure of women has a lot of effect on decision-making (Lodha 2000). Number of studies conducted in third world countries revealed significant variations about women's economic activities outside the home. With the exception of Malaysia, Women in South – East Asia enjoyed relative freedom to engage in and out of house economic activities.

Kallgren (1977) observed that in South Asia female non-familial activities are restricted by cultural constraints of female mobility and female seclusion. However, according to Durand (1975) Women's participation is greater in agriculture than in the non-agriculture sectors. A similar trend was also observed in Bhutia society of Sikkim in the present study. A study in Pakistan also indicated that approximately 73 per cent of women belonging to agricultural household were in labour force. Among them 54 per cent were engaged in agricultural work on their family farms (United Nations 1993). In another study, Boserup (1970) observed that the proportion of women as agricultural wage labourers was 44 per cent in India, 40 per cent in Sri Lanka, 39 per cent in Malaysia and 35 per cent in Thailand. Palileo (1995) observed that the proportion of female working population engaged in agricultural and related activities ranged from over 50 per cent in Korea and Sri Lanka to 80 per cent in India, Pakistan, Thailand, Malaysia and Nepal. Thus the nature and magnitude varies from region to region and even within a region their participation changes according to social hierarchies, farming pattern, socio- economic status and even ecological zones (Swaminthan 1985). In the hill areas tribal women are much heavily burdened. Hutton (1922) described how among the Lhota Nagas in Nagaland, women work side by side with men. However, among them in the field as well as at home work is divided between the sexes. Lama (1994) notes that among the Lepcha and Bhutia household, heavier works are often performed by men except ploughing, women are actively involved in agriculture. Thus there is hardly any activity which women do not perform and hence there is a relationship between work structure of Bhutia women and its impact on her position in her decision – making role especially in Sikkim.

It was observed that Bhutia women in Sikkim participated in various activities outside the house, like attending to farm labour, horticulture, post-harvest operations and tending domestic animals besides their household chores. In the traditional Bhutia society, women shared agricultural work equally with men but development of new social conditions has changed their working conditions. It was found that Bhutia women work in Government services, as Tradesmen or in Handicrafts / Handloom sector. The level of education is indicative of the kind of occupation the Bhutia women perused The educated Bhutia/Lepcha are getting more and more interested to join administrative services. Even the young Bhutia/Lepcha women with peasant backgrounds are migrating to urban area to seek white-collar jobs in government, semi- government and other offices (Bhattacharya 1994). Women who are relatively less educated are confined to household chores and also empirically, engaged in agriculture. It was found that only 15 per cent women are engaged in government jobs. Their low level of education is a factor that conditions the lesser representation of women in occupations outside home and their maximum concentration in agriculture. It was found that women who had graduation degrees or above were generally employed. Bhutia women's participation outside their homes has increased due to increasing urbanization, which is reflected from the opportunities that are available to them.

The data [Table 1] revealed that in Sikkim, agriculture was the main occupational opportunity available to the Bhutia women. This was mainly due to the fact that agriculture was widely practiced In Gangtok women had some degree of occupational diversification, whereas in case of women in other areas opportunities remained limited. However, women's twin role of housekeeper and also a contributor to family income restricted their mobility in and around the village as they persuaded these opportunities very close to their residential places. Along with the availability of employment opportunities, the analysis of nature of work participation of the respondents was also done. The data [Table 2] indicates that majority of the women were involved in the household, agricultural and horticultural activities.

Table 1
Employment opportunities

Employment	*Number*	*Per centage*
No response	1	0.5
Agriculture	120	60
Trade	20	10
Government service	50	25
Handicraft/handlooms	3	1.5
Others	6	3

Table 2
Nature of work participation

Nature of work	*Number*	*Per centage*
Govt. service and household	40	20
Household, agriculture, horticulture	60	34
Household, agriculture, tradesmen	68	30
Student and household	10	5
Household	22	11

The number of hours they spent on carrying their household activities revealed that most of the women are hard working. The data [Table 3] indicates that 63 per cent women, on an average spend 5 hours in household and 25 per cent women spent 6 hours in meeting their household needs. The number of women spending more than eight hours is only 6 per cent. Bhutia women after spending time in the household activities, utilize the remaining time for other income supplementing activities. In comparison to data in the above table the data in table [Table 4)] reveal that in addition to their household work 50 per cent of the Bhutia women spent an average of 5 hours in work other than the household activities.

Table 3
Average hours of work in the household

Hours of work	*Number*	*Per centage*
Not applicable	6	3
3	0	0
4	6	3
5	126	63
6	50	25
8	12	6

Table 4
Average number of hours spent on work other than household

Hours of work	*Number*	*Per centage*
Not applicable	6	3
3	3	1.5
4	5	2.5
5	100	50
6	76	38
8	10	5

The 38 per cent women who spent 6 hours in other than household activities follow them. The number of women spending 3 hours is only 1.5 per cent, which means that Bhutia women in Sikkim are doubly burdened. Similar results were also reported by Rath et al. (1988) in a study of tribal women in Orissa where out of the total 16.50 working hours per day of women, 7.50 hours were allocated to field and forest related activities. Lapche (1987) reported similar results about the farmwomen of Sikkim. Women's work participation also depends on the attitudes that men have towards their participation, because it is only when women have the consent from the males of their families, can they work without any pressures.

The table [Table 5] discerns that 20 per cent men do not appreciate women working outside and 30per cent generally did not have any problem with the women working. This indicated that generally Bhutia men in Sikkim appreciated women working outside.

Table 5

Men's attitude towards women's work

Attitude	*Number*	*Per centage (%)*
No appreciation	33	20
Appreciation	74	34
Do not say anything	93	30

The data with regard to women's occupation, their work participation and the attitude of men and women reveals that most of the women are found participating in work irrespective of the attitude of men. Bhutia men however, did not indicate a pessimistic attitude towards women's participation. Economic freedom provides women self – assurance and self-confidence, resulting in changes in their attitudes, aspiration and effective participation in different walks of life. Therefore, the impact of her work participation should be felt in their field of decision-making.

Decision-making

Decision making involves allocation of resources and distribution of roles within family, in other words, the extent to which women have equal voice in matters affecting themselves and their families, their control over material and other resources, the authority to make decisions, and freedom from constraints on physical mobility are included under decision making. These are also important indicators of autonomy. Autonomy has been defined as 'the degree of women's access to, and control over material resources (including food, income, land and other forms of wealth) and to social resources (including knowledge, power and prestige) within the family in the community, and in the society at large' (Dixon 1978). Decision-making is one component that reflects women's position in different areas of household, managements, economic participation, and access to resources (Upadhyaya 1991).

These alternate dimensions of autonomy have rarely been measured empirically and what have been routinely used for analyses are education, marriage age, profits of economic activity etc. which are regarded as inadequate for measuring dimensions of female autonomy. When a women works she increases her ability to sanction or control, because of the financial contribution she makes to the family she wants to have a say in family affairs and wants herself to be involved in decision making regarding children's education, shopping, budgeting or financial matters.

Information was also gathered concerning their autonomy and decision-making power within the household because the inclusion of these dimensions would allow a better understanding of the position in the various decision making processes. Four dimensions of autonomy were selected and indices for each were created. Economic decision-making authority and child related decision-making; mobility;

freedom from threat from husband; access to economic resources and extent of ownership right over material possession were taken into consideration.

Economic decision-making authority

Economic decision-making authority is represented by information on the participation in three economic decisions: the purchase of food, major household goods and jewellery. When women has not been or does not buy any asset she is given a score of 1 or low. If the woman has been consulted she is given a score of 2 or medium and if a woman has chosen these articles on her own she is given a score of 3 or high.

Children related decision

Children related decision making authority is represented by information on whether the women is the main decision maker on issues as what to do if a child falls ill, discipline of the children, how much to educate children and what type of school should children attend. The woman is given a score of 1 or low when she has no say at all, 2 or medium if she makes a decision on any two or three issues and she is given a score 3 or high if she is a major decision-makers in all four. A sample of 200 respondents was taken separately for this study.

Mobility

Five places – the health centre, the home of a relative or friend, the market, cinema and the neighboring village- are considered in terms of Mobility. She is given a score of 1 or low if she can go only if escorted, 2 or medium if she goes unescorted but needs permission for going to some of the places and if she can move about unescorted without permission to every place, she is given a score of 3 or high.

Freedom from threat

The index of freedom from threat ranges from 1 to 3. 1 or low is assigned to women who fear their husband and are beaten by them; 2 or medium if they fear and are not beaten; 3 or high if they neither fear nor suffer beating at the hands of their husband.

Access to financial resources

Access to financial resources is summed up in four questions: (a) having a say in how household income is spent; (b) getting cash to spend; (c) operating bank accounts; (d) being free to purchase small gifts or items of jewelry. Women is given a score 1 or low if she has no control over any of the resources and everything is looked after husband, in laws or parents, 2 or medium if she has access to some resources but jointly with husband, parents or laws and a score 3 or high if she has independent access.

Control of ownership over material possession

This includes questions like (a) whether any of the family's valuables (land/jewellery) belong to the woman (b) whether she can support herself through her own savings. This is measured with the help of giving score 1 or low if others (husband/in-laws) have ownership rights; 2 or medium if it is jointly owned and 3 or high if wife/respondents have sole right.

The above indicators of autonomy were used to analyze Bhutia women's position in society.

Economic decision-making is an important indicator of women's decision-making power in terms of their position in the household. Women's participation in work outside the house is conditioned by various regional and cultural factors and in most cases it is the compulsion of economic pressures that women go out to work. This fact has been acknowledged in a number of studies conducted on women's participation in work force in the developing countries (Niranjna 2000). Bajwa (1983) pointed out that woman's economic role affects their position in the family. It was apparent from the respondent's response that in the entire four district women had a medium level of control when it came to decision-making on their economic assets. It was observed that although Bhutia women had control over/decision on day today cooking/ procuring items but when it came to buying expensive items she did not have much say. The data [Table 6] shows that the women had medium position in access to economic assets in Sikkim.

Table 6
Bhutia women in decision-making (access to economic assets)

District	*Low (%)*	*Medium (%)*	*High (%)*
East	6	66	28
North	20	50	28
West	30	56	14
South	30	70	10

Acharya and Bennett (1982); Badigar and Rao (1980); Parthasarthy (1988) have documented that women wage earners have better status within the household because their economic contribution to the family has greater visibility and their independent earning gives them a higher bargaining power enhancing their status in terms of decision making power. This may also be applicable to the Bhutia women of Sikkim. After the analysis of the economic contribution, information on their locational awareness and freedom of mobility was sought. It was observed that the unmarried and married women were allowed to move alone with prior permission either from parents/husband: that means they had a medium degree of freedom when it came to their freedom of mobility. However the divorcees and the widows in the Bhutia society scored a high degree of freedom as they were not required to seek permission of anybody in the family, as they had no questioning authority in the family.

It was evident from the data [Table 7] taken, regarding the position of Bhutia women on the matter of domestic violence that more or less the Bhutia women neither feared nor were subjected to any violence. In general the result shows that in Bhutia society, the women believed in equality and that reflects in their day today living. They retaliate when victimized by their men and family and very seldom they kept quite out of fear. In case of very bad situation only, they were found to be reporting to the Panchayats. Ghai (1981) reported that women's ownership over productive resources increases their decision making power. One who controls the family property occupies an important position in the power hierarchy. Power over money is an important factor in decision-making control over finances reveals one's position within the household.

Table 7

Response of Bhutia women to freedom from threat

District	*Low (%)*	*Medium (%)*	*High (%)*
East	0	0	100
North	0	2	98
West	0	2	98
South	0	3	97

In Bhutia society it was generally observed [Table 8] that in the entire four districts women had a medium degree of hold over the financial matter in the family. Majority of the women could not draw money without the consent of their husband, which shows their lack of control in the family financial resources. Children centered decision making is also an important indicator which effects women's position and autonomy in the family.

Table 8
Role of Bhutia women in decision-making (financial matters)

District	*Low (%)*	*Medium (%)*	*High (%)*
East	30	46	24
North	22	58	20
West	30	50	20
South	28	52	20

In Bhutia, because of the patriarchal nature of the society, women were not free to take independent decision with regards to the children but they could make decision about the education of children. Therefore, it can be summed that she enjoys a medium position and generally took decision with concurrence of her husband. The result [Table 9] presented in all the tables confirms that Bhutia women have a medium level of decision-making authority.

Table 9
Role of Bhutia women in decision-making regarding children

District	*Low (%)*	*Medium (%)*	*High (%)*
East	16	48	36
North	20	62	18
West	34	60	6
South	36	60	4

Case Study

I

Tshering Wangchuk Bhutia* is 32 old married women and a resident of Tathangchen, Gangtok. Her husband works in one of the Government Offices in Gangtok. She is a housewife and spends her time looking after her children and her

*Names of the respondents have been changed.

household. She freely moves to the market to purchase articles for the household and is well aware of all the important places in Gangtok like Post Office, Hospital, and Police Station etc. She goes to pick up her children from school and also teaches them because she is also a dropout after class twelve. She has no idea how her husband makes investments because she never operates the bank but he does give her money to run the house. Her parents stay in West Sikkim so she goes once a year to visit them with her children. She has no desire to work because her husband earns very well. She is very repelled with the idea how husbands beat up their wives. Her Nepali neighbor faces this problem but never asks anyone for help, she tried to help her out with the intervention of women commission but the torture continues for this lady. She feels women's position in society is very much dependent upon the attitude of her parents before marriage and her husband after marriage.

II

Pema Lahmu is 28 years old, unmarried and a resident of Barfung. She works in her small field where they grow vegetables to sell in Rabong Bazar. She has had education till class 5 and because she was not a bright student she had to drop out from school. She has two elder brothers who are married and stay in Gangtok with their families. She stays with her parents and her younger brother. She spends 5 hours doing the household chores and devotes on an averages 6 hours in her own agricultural field. Her parents are very appreciative of her hard work and want that she should marry someone from Barfung so that they do not lose her labour. She is able to move freely in nearby areas, sometimes accompanied by her father or brother, but most of the time alone. Since she stays with her parents they look after most of the requirements she can cook food of her choice and even purchase articles for the household of her own but she can never buy valuables like jewellery. She has some money that she can spend on her own, but she never visits any bank or takes any interest in any of the financial matters. Her parents want to give her some share in family property because she is the one who is looking after them she commands a good position in the family.

III

Choden Bhutia is 68 years old widow and a resident of Singyung (West Sikkim). She has four children who are all married, but she lives with her eldest son who is a farmer in Singyung. She does not do any household chores because she is very old now and has also developed cataract in her eyes. Her daughter in-law treats her with respect and her son always gives her some money for her personal use. Earlier she was very fond of weaving carpets but now because of her eye problem she is not able to do so but once in a while she goes and sits in the shop that is jointly run by her middle son and his wife. She has visited Gyalshing, which is the nearest town but has no idea about any of the important places like Post Office, Public Health Center etc. she moves freely without permission to her neighbors place and also visits her sister who stays nearby. She has some traditional jewellery, which she wears for special occasions; she wants to keep it because she wants to give it to her daughter. The land has already been divided between her three sons when her husband was alive so she feels she should give something to her daughter. She regrets not having any formal education but otherwise she is happy with her position in the family.

IV

Rinzing Lhamu Bhutia is 35 years and a resident of West Sikkim. She was eighteen years old when she married this Nepali boy (Rai) from East Sikkim. Her was an inter community marriage as her parents refused to accept her. She moved to Gangtok with her husband who was a driver and her two children were born within a span of four years of her marriage. Her husband started inflicting abuses on her and started quarreling with her over small matters and finally one day he left her and started living with another girl. Rinzing took up a job in SITCO Watch factory and started looking after her children. In her extra time she makes hand knitted pullovers on order. She spends 5 to 6 hrs on her household work and approximately 6 to 7 hrs in the factory. She freely moves about and is aware of all the important locations of the society. She never asked her parents for any help but feels that she should have her husband punished. She freely spends money because she says she has earned it. She wants to give the best of education to her two children who are now studying in Standard III and IV respectively because she feels education is a must. Although she does not have any assets to boast about but she is happy with her life. She has her friend circle and feels that at some point of the life her parents will forgive her because that is one thing that she really misses.

Conclusion

There is a definite pattern to the kinds of decisions in which women participate and they are generally not threatening in patriarchal context. All the dimension of autonomy appears closely related and none of them are weak or less consistence. Bhutia women in almost every index of autonomy decision making authority, mobility, access to and control over resources, experience freedom as such. A highly gender- stratified culture was not seen and relations were equal. However, in most of the cases women did not take decision independently but always with their husband or in-laws/parents consent. Despite women working in economically gainful activities they enjoyed a moderate status in the economic as well as social affairs of the family. Where women did enjoyed a higher position it was because of a level of education. Education level, which is a cognitive component, can affect qualitatively the time spent on work structure and also increases awareness regarding access to resources. Acharya and Bennett (1982); Agarwal (1986); Kaur, (1987); Oberoi and Sandhu (1991) have also reported that women's involvement in decision making is associated with her educational level. With education and employment her role in total management has become very significant and they do not suffer from onerous and equitable position. The data with regard to their participation in household, agriculture, horticulture work indicates that the life of Bhutia women is burdened with these works. These women work on an average for 6 hours a day and these works supplement them in the household income and also earn her the recognition from her men and family. It is therefore quite obvious that women's work gives her good social standing and also the right in decision-making process of the family and in turn adds to their empowerment.

References

Acharya, M. and Benett, L. (1982). 'Women and the Subsistence Sector Economic Participation and Household Decision-making in Nepal'. *World Bank Staff Working Paper No. 526,* World Bank, Washington DC.

Agarwal, B. (1986). 'Women Poverty and Agriculture Growth in India'. *Indian Journal of Peasant Studies*. 13(4), pp.165-200.

Badigar, C.L. and Roa, M.K.S. (1980). 'Participation of Farm Women in Decision-making on Farm and Home Aspects'. *World Agricultural Economic and Rural Sociology Abstract*. 23(9), p.740.

Bajwa, A. (1984). 'Decision Making by Rural Women in Selected Household Activities'. Unpublished Master's thesis, Punjab Agricultural University, Ludhiana.

Balikci, A. (2002). 'Buddhism and Shamanism in Village Sikkim'. London School of Oriental and African Studies, Unpublished Thesis.

Bhattacharya, A. (1994). 'The Bhutia-Lepcha Women in Sikkim: Tradition and Response to Change' in Lama M.P (Ed) *Sikkim: Society, Polity, Economy, Environment*. Indus Publishing Company, New Delhi.

Boserup, E. (1970). *Women's Role in Economic Development*. George Allen and Unwin, New York.

Dixon, R. (1978). *Rural Women at Work Strategies for Development in South Asia*. John Hopkins University Press, Baltimore.

Durand. (1975). 'Rural Development and Women's Economic Roles in Asia'. in Das M.S. and V.K. Gupta (Eds) *Social Status of Women in Developing Countries*. M.D. Publication, New Delhi.

Engles, F. (1948). *The Origin of Family, Private Property and the State*. Moscow Progress Publishers, Moscow.

Ghai, D. (1984). 'Resource, Power and Women'. *Proceedings of the African and Asian Inter- Regional Workshop on Strategies for Improving the Employment Condition of Rural Women*. International Labour Organization, Geneva.

Giele J.Z and Smock, A.C. (1977). *Comparative Perspective on Women*. John Wiley and Sons, New York.

Hutton, J.H. (1921). *The Sema Naga*. Macmillan, London.

Kallgren. (1977). 'Rural Development and Women's Economic Roles in Asia'. in M.S. Das and V.K. Gupta (Eds) *Social Status of Women in Developing Countries*. M.D. Publication New Delhi.

Kaur, S. (1986). 'Role of Women Rural Development: A Case Study of Haryana'. PhD Thesis, Hisar: Haryana Agriculture University.

Kothari, C. R. (2004). *Research Methodology, Method and Techniques*, New Age International Publishers, New Delhi.

Lall J.S (1981), *The Himalaya: Aspects of Change*. New Delhi: Oxford University Press, 1981

Lama, M.P. (1994). *Sikkim- Society, Polity, Economy, Environment*. Indus Publishing Company, New Delhi.

Lapche, T.T. (1987). 'A Study of Farm Women's Contribution to Agricultural Production in East District of Sikkim State'. Master's Thesis, Gujarat Agricultural University, Anand, India.

Lodha, N. (2000). *Status of Tribal Women: Work Participation and Decision Making Role in Tribal Society*. Mangal Deep Publication, Jaipur.

Marx, K. (1963). 'Economic and Philosophical Manuscripts'. in T.B Bottomore (ed), *Marx Early Writings*. Penguin, Harmondsworth.

Niranjna. (2000). *Status of Women and Family Welfare*. Kanishka Publishers, New Delhi.

Oberoi, K and Sandhu, M. (1991). 'Status of Women in the Society in Present Situation'. In Puni, R.K. (ed.) *Women in Agriculture: Their Status and Role*. Vol.1, Northern Book Centre, New Delhi.

Palileo, G.J. (1995). 'Rural Development and Women's Economics Roles'. in Asia. In Das, M.S. and Gupta, V.K. (eds) *Social Status of Women in Developing Countries*, Deep and Deep Publication, New Delhi.

Parthasarthy, G. (1988). 'Public Policy and Trends in Women in Development', Paper Presented at Workshop on *Women in Agriculture*, Centre for Development Studies, Trivandrum, February 15-17.

Rath and Das. (1988). *Rural Women at Work*. John Hopkins University Press, Baltimore.

Rothschild, S.C. (1970). 'A Cross-cultural Examination of Women's Marital Educational and Occupational Options'. *Acta Sociologic*. 14, pp.95-113.

Rodman, H. (1972). 'Marital Power and the Theory of Resources in Cultural Context'. *Journal of Comparative Family Studies*. 3 Spring, 1, pp.50-69.

Sherring, C. (1906). *A Western Tibet and the British Borderland*. Edward Arnold, London.

Singh, K.S. (1993). *People of India: Sikkim.* Volume XXX, Anthropological Survey of India, Seagull books, Calcutta.

Strauss, L. (1953). 'Social Structure', in A. L. Kroeber, (ed) *Anthropology Today.*

Swaminathan, M.S. (1985). *Imparting a Rural Women User Perspective to Agricultural Research and Development.* IRRF, Philippines.

Upadhayay, H.C.(1991). *Status of Women in India.* Vol. 1, Anmol Publication, New Delhi.

Anonymous. (1993). *United Nations Women's Education and Fertility Behaviour: Case Study of Rural Maharashtra, India,* Department of Economic and Social Information and Policy Analysis, New York.

12

Cloth, Women and Social Change: Situating Handloom Weaving in Manipur

Otojit Kshetrimayum

Introduction

This paper tries to assess the culture of weaving in Manipur and examine the factors responsible for the development of weaving from a traditional craft to a commercial industry. Weaving in Manipur was started with an emotional attachment since the women and girls wove only for their near and dear ones and in each of their products, they tried to express something of personal emotion and communicate warm feeling through their fabrics that were intended to be worn. Weaving was done with a feeling to please the users and every product was, not only an article of utility to protect the body but also a memento and a medium for communication of feelings between relatives, friends or lovers. Women reflected their personality through their weaving. Therefore, by such reasons in almost Manipuri household, there were at least one or two looms i.e., loin loom. Every Manipuri adult woman wove for herself and her domestic consumption. The mother acted as the master-craftsman in the family. It became her duty to make her children learn to do their respective work. Cotton was used as raw material. It was grown and harvested by the males and the females did spinning and weaving.

Traditional Dress and Social Messages

Social messages sent by clothing, accessories, and decorations can involve social status, occupation, ethnic and religious affiliation, marital status and sexual availability, etc. Humans must know the code in order to recognize the message transmitted. If different groups read the same item of clothing or decoration with different meanings, the wearer may provoke unanticipated responses. In many regions of the world, national costumes and styles in clothing and ornament declare membership of a certain village, caste, religion, etc. A Scotsman declares his clan with his tartan. A Muslim woman might wear a *hijab* to express her religion. A male Sikh may display his religious affiliation by wearing a turban and other traditional clothing. A French peasant woman may identify her village with her cap or coif. Clothes can also proclaim dissent from cultural norms and mainstream beliefs, as well as personal independence. Religious clothing might be considered a special case of occupational clothing. Sometimes it is worn only during the performance of religious ceremonies. However, it may also be worn everyday as a marker for special religious status. For example, Jains wear unstitched cloth pieces when performing religious ceremonies. The unstitched cloth signifies unified and complete devotion to the task at hand, with no digression. The cleanliness of religious dresses in Eastern Religions like Hinduism, Buddhism and Jainism is of paramount importance, which indicates purity. There is also a cultural significance in the traditional dress of the Meiteis. The values of morality, aesthetic and ideology of the cultural history of the Meitei are expressed in terms of the colour combination, pattern, design and texture of fabric. There is also the introduction of foreign fabric as symbol of victory in representing the greater Meitei kingdom. Because of these cultural connotations to weaving, it was a very popular household industry in the traditional Meitei world. The change in this sustained tradition of the significance for weaving in Manipur, particularly among the Meitei, is that the art of weaving in its contemporary form had become commercialized.

Symbol and talisman

Each aspect of Meitei life, either of cultural value or of social establishment, has been well reflected through material culture. There is no material culture comparable with the traditional costume with respect to the establishing of ethnic identity, social hierarchy, aesthetic attainment and religious concept rooted in the cultural ethos of any ethnic group (Khelchandra 1998). Textiles are associated with the social and are a ritualistic event from very early times. Textiles' weaving seems to have played a pivotal role in the social and the economic life of the Indian people throughout the age. The continued existence of this craft helped in decentralization of the social and the economic power and provided with enough employment to the people. It is even truer, that this craft in the predominantly agricultural society is pivotal. The art of weaving made the village society a self-contained one, which is a characteristic of Manipur society too. The social functioning of Manipur is based on the code of personal relations and duties handed down from generation to generation. The people of this land do not take the execution of this craft as a compulsion but a sacred duty.

In Meitei lore, the weaving of textile appears as a replica of the cosmic process, during the dance of the creation, the male and female messengers of God (*Maiba* and *Maibi*) stimulate the sound of spinning and weaving. In other words, creation is a magnificent process of weaving to the cosmic being. It is a matter of great significance that in the traditional belief system of the Meiteis, it is the God of Handicrafts (Leisembi) who taught the Meiteis, the method of textile weaving as well as the mystique of the dance creation. There is, however, an alternative stand in the belief system of the Meiteis. According to this, it is not by the grace of God that man has learnt textile weaving; but it is by his own enterprise that man has picked up the skill from nature. He observed the spiders weaving their webs on the shrubs and bushes and he responded to this challenge.

Among the rulers, who are believed to have been great patrons of the craft in their time, mention may be made of Nongda Lairel Pakhangba (who is supposed to have ruled between 33 AD and 153 AD) and the king Loyumba (1074-1122 AD). The hierarchy in the social pyramid was then expressed through the products of the female craftsmen. Consequently, instead of elaborate paraphernalia of regal grandeur, the colour, size, shape of the simple wearing of apparel and ornaments were used to signify the status of the chief and other dignitaries. This became an important factor in the development of crafts work in the society. To differentiate the chief from the serf, the craftsmen (rather craftswomen in this context) had to produce goods of superior appearance and durability in the colour, pattern and workmanship. The craftsmen were always on the attempt to improve their skill and they had even to experiment with vegetable dyes and different threads viz., cotton, silk etc., to attain excellence (Haobam 1991:207).

Since early days, the Meitei Kings and the nobles offered clothes to *Lainingthou-Lairemma* (God-Goddess) for the prosperity of the population and the land. Such offered clothes were then received with prayers and worn by the mortals so that long life with generations of descendants could be ensured. The Meiteis of Manipur also believed that wearing of shirts sewn from the used *phanek*- women's loin dress of one's mother and grandmother could ensure victory in hunting and battles. It is believed that wearing of amulets containing within them a piece of their mother's *phanek* would be a good omen and can help them prevent from meeting evil spirits. There are very old clothes, which were presented to deities and carefully kept at the shrines of *Umanglais* or at the house of their owners. Believing these long-preserved clothes to be divine spirits and considering the water from these to be helpful in curing diseases, people worship them and collect the *machum*- water. In case such clothes are torn in some areas, these are usually not stitched. Instead, if at all necessary, stitching is done on a new cloth. Such kind of stitching is not done by the young people but by the old. It is because of the belief that stitching of such clothes shortens one's life. Thus the older people who are on the threshold of their death take care of these clothes and hand these over to their wards as invaluable treasures (Bahadur 1997:13-15).

While stitching or weaving a cloth, it was customary to leave a part uncompleted where it would be hardly noticed. It was because of the belief that there existed the Almighty, the Great Master and doing His work with perfection was considered to be a sign of egoism. If a woman wants to wear a *phanek* having a pattern known as *Akoibi-Mayek* then it should have only a few definite, limited designs. Otherwise it is believed that its wearer would become a widow. In early times, when parents arranged the dresses of their daughters, they used to arrange different types of dresses depending upon the ages of their daughters. Only widows used to wear a type of *phanek* having a certain pattern known as *Kumchingbi Hijam Mayek*.

Different colours and patterns of *phanek* worn by women signify their different *salais*. This is shown as follows-

Colour and design of the phanek	Salai/clan
• Lotus colour and kuhi-kum with red strips	Ningthouja
• Langhou phanek with black and white	Angom
• Blue phanek with neem machu and black strips	Luwang
• Kumchingbi with fine strips over the black	Khuman
• Yellow phanek with reddish strips	Moirang
• Kuman of Chigonglei colour with black strips	Khaba-Nganba
• Loirang phanek with reddish brown and white strips	Chenglei

In Manipur, there are types of embroidery, which are specially made for warriors and to be presented by the king as a mark of distinction:

- *Lanphee:* War cloth, it is a special type of shawl embroidered by the women at home, warriors use it at the time of going out for war.
- *Ningthouphee:* Waistcoat, which is presented by the king to the warriors of the country.
- *Saijounba:* Long coat, these were prepared with special embroideries for the very trusted courtiers of the king.
- *Phirananba:* Small flags delicately embroidered and used by the warriors as plumes on their turbans.
- *Namthang-khut-hut:* A design derived out of the head of Pakhangba on the wrapper, meant to be used only by the ladies of the royal family.

- *Khamenchatpa:* It is a ceremonial dhoti, which cannot be worn by any common person. It was presented as an item of reward by the king to brave persons.
- *Phiranji:* A red coloured blanket presented to the persons of merit. The colour of the blanket is a full red that signifies it as coming from the blood with the placenta.

Some of the other important designs are:

- *Temple design or Moirang phee:* This design is actually in red, green, black and blue colours and is woven in the borders of chaddars and sari. The princess Thoibi of Moirang is believed to have created this design.
- *Shamilanmi design:* The Shamilami design is embroidered with motifs of horses, elephants and butterflies etc. shades of orange, green, blue and red are very common in these designed shawls. The shawls with this design are popularly known as the Naga chaddar as the Angami Nagas used the design. They are presented to the braves.
- *Leirum:* This cloth thus woven however is used by the Meities in ceremonial occasions. Among the Meities it is an essential item of presentation during the marriage.
- *Chum design:* Cloth designed after pattern is exclusively meant for a presentation to be made by the king to the people of high proficiency in dance and song. Pundits and heroes also receive this cloth as recognition of their deep knowledge and heroism. This design is now woven not for the same purpose as in the beginning but on the ordinary door curtains, cushion cover and mats, etc. The shades commonly found are orange, brown, blue and chocolate.

The teaching of moral education through the medium of dress culture is also reflected by the styles of loin dressing, particularly the dhoti types adopted by the Meiteis on occasion of attending to the royalty. Thus, in the style of attiring in the *Kaptreng Thougal Phiset* style the loosely hanging tuft end to the front was gathered and carried to the back so that it did not hang freely in the front as it was considered irresponsible and disrespect to do so. For the same moral reason also in the mode of wearing dhoti, called *Thoukal Phiset Kunja Yeppi*, the front tuft end was picked up and tucked in at the waist on the right side. The *Kunja Kaobi* style of wearing dhoti was reserved for the king. No other male member, whatever level of social standing he might be, could wear the dhoti in this fashion before the king.

For the Meitei women, there are modes of wearing loincloth identifying the different status of the married and unmarried women. In the case of married women, the upper end of the cloth was raised up to cover the breasts. Unmarried women, on the other hand, put on loincloth around the waist leaving the upper part of the body to be covered by shirt. The married women's mode of wearing phanek is known as *Phidol Chingkhatpa* while that of a girl is known as *Phanek Koikha Setpa*. This marks the difference in status between womanhood and maidenhood. It was considered highly immoral for a girl to wear phanek in the married woman's mode and vice-versa, as it involved disordering of the structural arrangement of keeping these two categories of females in two different social statuses. The women had to take every possible care in public gatherings, particularly in the ritualistic occasion, that their phanek did not come loose. To avoid such awkward scenes, the Meitei women put on a shorter over-garment (locally called *Kanap*) around the waist in the form of a belt. Besides serving as a belt, the use of the garment over the inner loincloth was encouraged with an aesthetic sense. This dressing was prevalent among the married women, and in the case of the unmarried girls this was called *Khwandchet*.

The moral idea of simplicity and humility in the dress culture of the people was clearly visible when they were attired in plain white cloth on ceremonial occasions, particularly mortuary rites and ritualistic worships. On these occasions, the women put on *Pumngou Phanek* as their customary loincloth. Pumngou Phanek worn by the women on these occasions were of saffron colour with no border strips. The menfolk wore white in ritualistic occasions. The matter of moral value is also depicted in the distribution of the stripe-patterned phanek. This is again connected with the difference in cultural value attached to the colours of the phanek. The brighter colours are prescribed for the youngsters while the fainted ones are for the aged. Only the elderly women wore the phanek today called the *Kumjingbi*. On occasion of marriage, the elderly women put on any suitable striped phanek (Devi 1998:89-91).

Hill-plain relationships

The use of cotton fibre among the tribes of Manipur had been reported from the early days of their settlement. Since the reign of king Khagemba (1597-1652 A.D.), some tribes such as the Tangkhul, Kabui, Maring, Thangal etc., paid compulsorily a *Pon*-ful of cotton and *Kumna* (a vegetable dye from which indigo colour is extracted) as annual tribute to the Meitei kings. A good deal of the cotton raised by the inhabitants of Manipur Hill territory, which seems of excellent quality, finds its way into the bazaars of Manipur, there being no cotton grown in the valley. Trade, from the scanty nature of the hill productions not sufficient for the sustenance of the people, is confined almost entirely to the bartering of raw cotton and a few other articles in the bazaars; salt is chiefly taken in return (Brown 1975: 18-19). The Nagas, especially those to the north, often make large purchases of cloths in the Golden Bazar, both of ordinary white with red and blue stripes, and also the more special cloths of a dark-blue ground with figures of animals woven in red thread round the borders (Hodson 2003: 27). The *Shamilanmi* design is embroidered with motifs of horses, elephants and butterflies etc. Shades of orange, green, blue and red are very common in these designed shawls. The shawls with this design are popularly known as the Naga chaddar as the Angami Nagas used the design. The Bakey design was named after a Kuki woman who invented this design. This design was a tribal design and was later introduced to the center. This design was woven originally for skirts but now this design is practically found on any type of shawl and bed covers. The common shades are blue, green and maroon. There is also *Kudam Manbi* design, which is known by several names. The most common one is, button design but the other popular one is known as Mitlaobi design or broad-eye design. The idea of this design is said to have originated from the patterns made of beads and cowries by the Kukis.

The Mao weavers have produced a cloth called *Khupidekhusa* having needle works. It is a cloth, which imitates the *Sami-Lanmi* made by Meitei weavers, but meant for tribal leaders, which bears conventional figurines. At the time of merrymaking of the Meitei deity, Wangbren the Anal offer cloth at its temple. It is because of their belief that an Anal girl was married to Wangbren. So they consider it their duty to gift a cloth at this function in the name of the Anal girl.

The Craft of Weaving: Mythical Evolution

There are many myths about the origin of cloth and weaving in India. The hill people of east and west India tell a story of the first weaver, a girl named *Hambrumai*, who used to lie in the forest and look up at the interlacing of the branches and leaves, or watch the ripples in the stream and then imitate these images by weaving beautiful cloth. However, the porcupine, *Harium*, was jealous and tried to steal her cloth. He shattered her loom with a rock and let the fragments fall into the river where they washed downstream to the plains. There, the people pieced the loom back together and learned how to weave.

The art of weaving of the Meiteis of Manipur is based on a mythical foundation. According to the manuscript on mythology called *Leinunglon*, the supreme goddess of the traditional pantheon, *Leimarel* is said to have had introduced the work of weaving as a necessary item of work in the whole course of creation of the universe. It is also stated that the goddess *Panthoibi* was the first primeval celestial ancestress of the Meitei and it was she who introduced the art of weaving by imitating the pattern of weave that a spider applies when it is weaving its cobweb. The *Leinunglon* mentions that the first cloth made by Panthoibi was of material obtained from the soft membrane of wood and bamboo. It is recorded in the same source that *Tamitnu*, one of the seven primeval celestial ancestress of the Meitei, produced cotton for manufacturing cloth for the first time. For this, She is known as the goddess of cotton. The manuscript also states that Leimarel, the chief of those seven primeval ancestresses, invented the art of making thread yarn from raw cotton for use in weaving. To commemorate this achievement, the office of *Sinkhombi* was instituted, and maintained throughout the history of the people, to look after the craft of weaving. The place where the weaving is done was called *Sinsang*. The word sinsang also refers to the institution that functioned as the administrative machinery that dealt with the production of royal cloths.

According to the Meitei mythological traditions, the most ancient female dress is the *Chinphi Phanek* and the *Laiphi* (muslin) and the first male style of dressing the loin is the *Khwangli Laikhal*. The creation myth of the Meiteis holds it that the seven goddesses and the nine gods who performed the creation works were clad in the respective dress mentioned above.

From times of yore, the Meiteis have been observing the ritualistic festival of *Umanglai Haraoba* (or Lai Haraoba). One of the important rituals of Lai Haraoba is the performance of the dance of *Pam Yanba*. In this ritual dance, the hand movements of the *maibis* (priestess) accompanied with songs depict cultivation of land and sowing of cotton seeds, plucking of cotton, ginning, carding, then spinning and weaving of clothes, cutting it and finally dedicating it to *Lainingthou* (male deity) and *Lairembi* (female deity) for the prosperity and well-being of the people and the land. From this ritual dance depiction, it can be asserted that from early times people dedicated their hand-woven clothes before using to Liningthou and Lairembi. Moreover, amongst the Meiteis, use of hand spun and hand woven cotton fabric is part of their traditional ritual. For instance, not only the traditional priests and priestesses (maibas and maibis) use hand spun and hand-woven fabrics in the performance of Lai Haraoba but also those who participate in the procession.

Weaving Under the Royal Patronage

Patronising the Craft

A factor of great importance was the patronage extended by rulers of Manipur to one craft. It is, however, not clear whether the active force for this patronage was always derived from the aesthetic inclinations of the rulers, or whether the patronage was a dimension of the shifts in favour enjoyed by the different ethnic groups. Besides, much of the patronage was an investment for reaping benefits out of a flourishing industry is also not clear (Bahadur 1997: 11). Among the rulers who are believed to have been great patrons of the craft in their time are highlighted in this section.

The reign of Nongda Lairen Pakhangba, who ascended the throne in 33 A.D., according to *Cheitharol Kumbaba*, the royal chronicle of the Meitei kings, is described in the ancient manuscripts of the Meiteis to be well advanced in dress culture. The sacred manuscript *Meitei Puya Wari Machang* describes the ordinary dress of Nongda Lairen Pakhangba and his consort Laisna as Maimu Tonsen, a fashion of wearing turban for the king, and *Chinphi phanek*, *Langla Khaothang Samba* style of shirting and *Sarong Namthang* (over garment wrapped over the loin) for the latter. According to the puyas, *Pakhangba Phambal* and *Kanglei Sanglen Puba*, the coronation robes of Nongda Lairen Pakhangba and his queen Laisna were as follows:

Dress of the King	*Dress of the Queen*
Lollei Kokyet	(headdress) *Luren Luhao* (a style of hair keeping in which the hair is tied into three knots, and decorated with hornbill feathers)
Phiren Phurit (white shirt)	*Kumlang Phanek* (loin dress of black colour with narrow yellow or white horizontal stripes)
Chari Ningkham (a kind of loin overgarment)	*Phiren Kanak* (loin overgarment of white colour)
Konkha Phurit (sleeveless gown)	*Kabrang Kanak* (a band of cloth in silk to cover the breast)
Samjin	*Khaon Phurit*

The reign of Khui Tompok witnessed remarkable achievement in the field of arts and culture, with particular reference to dress. It is said that items of dress, like the *Khamen Chatpa* and the *Ningkham Samjin* were introduced for the first time during his time. There is also an account of the introduction of Ningthouphi[1] during his rule. Another historic event during the time of Khui Tompok was the introduction of the shawl Leirum.[2] His reign is also credited with the introduction of sericulture for the first time.[3] There was an official called *Phisahanjaba*, who looked after the weaving of cloths and their designs and colour pattern.

The manuscript *Ningthouphi Saba* states that the invention of a second type of Ningthouphi was made during the reign of Meidingu Naophongba. The dress so produced was more developed than the former. The reign of the next king, Naothingkhong, marked the introduction of such coronation dress like the *Phingang Pheijom, Innaphi Langdom, Phiren Phurit, Konkha Phurit, Pakloi, kokyet*, etc. Sources such as the *Naothingkhong Phambal Kaba, Meitei Ariba Laining Laichat, Phambalon*, etc., stand to add evidence to this (Devi 1998:5). The artistic decoration of the Meitei dress, particularly the dress worn by women with embroidery work, first appeared during the reign of Yanglou Keiphaba (969-984). The historic landmark in the history of

---

[1] The name of the weaver who wove the particular type of Ningthouphi called Waphak Phurit is recorded in the written account Ningthouphi Saba (a text o the art of weaving royal dress) as Laikokshu. According to this source, Laikokshu wove the Ningthouphi with designs adopted from patters o the dorsal side of the Hangkok (chameleon). It records that the historic event of manufacturing this royal cloth was started on a full moon day, a Saturday in the month of Mera Tha, the seventh month in the Meitei lunar calendar, roughly corresponding to the English month of November and that the work was completed in seven months.

[2] Originally, leirum was a cloth of the Angom community. Such shawl locally named Charei Paroitheg Malou Koya Yekpa of ride's family was presented by the parents of ogmoiu Ahogi to uga Thayaga, a Meitei ole who prayed the hands of ogmoiu Ahogi for Khui Tompok as mark of respect and acceptance. This occasion of presenting Leirum from bridal family in relation with the marriage of Ahongi to Tompok has been give much cultural importance, so much so that throughout history, and till date, in Meitei marriages the gifting of Leirum cloth is an unavoidable item for the bridal gifts.

[3] The Meitei culture induced a mythical story to justify the point. Nongmeinu Ahongbi, the queen of Khui Tompok, gave birth to a daughter who did not survive long. The royal couple was shocked at the premature death of their daughter. After some time they found to their surprise mulberry worms appearing on the grave of their deceased daughter.

the art of dress of the people during that time was the introduction of the embroidery works of *khoi* pattern on the border of the female loindress *Mayek Naibi* (the stripped female lower garment) (*ibid:* 6).

The reign of Meidingu Loiyumba (1074-1122 AD) was marked by the achievement in producing a great variety of dress, and during this period an organized administration of weaving industry was introduced. The weaving culture flourished during his rule, as is suggested by the recording of names of royal weavers such as Naotam Phisaba, Ngangdi Phisaba, Yangnu Phisaba, Heisu Naha Phisaba and Khiriyam Phisaba in the manuscript *Loyumba Shilyen*. He also introduced a new design of the Ningthouphi by entrusting his head weaver Roso to depict on the cloth the design of the *Urit Mahum* (the nest of a local bird). The new colour of thread, namely *Makong Meiri* (flame red) and *Amuba* (black) were used in producing the high and low relief of the pattern of the nest on *Muga Machu* (light yellow) background.

Kwakpa Thawanthaba, son of Meidingu Thangbi Langthaba (1302-1324 AD) was considered as the patron of introducing Hindu custom. Because there were possibility that the male fashion of wearing *pheijom* (male loindress) was more influenced from the West by people who had been exposed to the influences of the Aryan culture. There are historical evidences recorded in the local literary sources to point to the contact of the then Meitei people with the people with those people of the West.[4]

The reign of Meidingu Kongyamba (1324-1335 AD) marked an achievement of Meitei costume by adding all attires of the Khuman after defeating the Khuman king Lamyay Kaikhinba. Later on, Meidingu Ningthoukhomba (1432-1467 AD) added the Moirang dressing style by conquering Moirang Ninghtou Sana Hongba. The reign of

[4] Here the most telling historical information is that Kwakpa Thawanthaba married a princess of the then Mleecha dynasty of Kamrup (modern Assam), named Chingurembi.

Meidingu Kiyamba (1467-1508 AD) witnessed the introduction of new items of dress in Manipur, originating from the West as well as from the East of the region. The fashion of putting on dhoti, called *Kaptreng Thougal Phiset*, was introduced for the first time in this period along with the advent of the cultural practice of worshipping the Hindu god Vishnu. Among the many new cultural things coming from the Shans, the female loincloth *Kanap* is an important specimen.

The production of silk and other cloths decorated with embroidery works in floss silk and the use of these cloths became very popular during the time of Meidingu Khagemba (1597-1652 AD). Besides the introduction of new items of culture, particularly dress and dressing styles, those traits and complexes that had been prevailing in the Meitei culture became enriched with the new additions and improvements wherever necessary in respect of the technique of making cloths, textile design, decorative artistic works and motifs. Khagemba enhanced production of cloths. The King conceived the practice of gifting specific award cloths (known as Mana Phi) to his talented subjects in various fields. It was also known as *Pari Phi*.[5] It is held that Khagemba introduced the offer of Wanphak Phurit as gallantry award. From the report on the popular use of the Maring (one of the tribes of Manipur) cloths by the Meitei people in his time, the introduction as well as the popular use of the Maring cloth can be considered as the follow-up of the conquest of the Maring.[6]

Mentioned may be made of the elaborate system of arrangement of the position and status of the nobles in

---

[5] There is a historical occasion for the naming of this cloth. In the days before the reign of Khagemba, the talent award was given in the form of arable land in terms of unit of measure of paddy, called *Pari*. During the reign of Khagemba, this form of giving award was replaced with cloths. However, the name of the earlier system continued in practice as in naming the new form; hence the name Pari Phi.

[6] *Cheitharol Kumbaba*, the Royal Chronicle, records that Khagemba fought a successful battle with the Maring tribe in the year 1630 AD.

Khagemba's court; the horizontal as well as the vertical arrangement of the position of his nobles and the other public men as well, were expressed in terms of different dress and modes of wearing the dress. It is also recorded that he prescribed dress: loincloths, turbans and shirt in black for his army (*Ibid:* 15-16).

'Feudalisation' of Weaving Craft

'Feudalisation' here refers to the process whereby the king took control over the economic occupations of the artisans and assigned occupations to each lineage group and also sex role differentiation and social exclusion of the artisans. In traditional Indian society, one's occupation is a function of caste. Within India, textile production is under the domain of certain castes or sub-castes, and even aspects of their use are governed by caste.[7] However, in Manipur, a family becomes an occupational unit by virtue of the fact that the king has assigned a particular occupation. For instance, a particular job was assigned to a *yumnak* (lineage). Thus the names of each yumnak would normally indicate the role assigned to them traditionally. Nearly every household is capable of weaving all the cloths needed by her family, and the simple loom stands in the verandah of the house. These cloths are mainly intended for wear and not for decorative purposes (Hodson 2003: 27). However, the association of lineages with productive labour processes was formally institutionalized in the early twelfth century during king Loiyumba's reign (1074-1112 A. D.). The 'feudalisation' of

---

7 Saliya (or Chaliyan or Sali or Sale) is an erstwhile Malayali weavers' caste found mostly in Northern Kerala and Southern coastal Karnataka. The main weavers' castes of South India are Sale, Devanga and Kaikkolar. Among these the first two castes are native to Kannada and Telugu speaking regions by inscriptional evidences. Whereas Kaikkolar community is native to Tamil speaking region, Padmashali is a Telugu weavers' caste or social group found largely in the state of Andhra Pradesh in India. Ramdasia Sikh is a caste in India. Members of this community were engaged in the profession of weaving. They may at times be also referred to as *julaha* meaning a weaver in Punjabi and Hindi. They are found in the states of Punjab, Haryana, Delhi, Rajasthan, Uttar Pradesh, Uttarakhand and Himachal Pradesh.

handicrafts and industries and king's control over the economic occupations of the artisan is clearly embodied in the royal decree *Loiyumba Shilyen* issued by the king himself in 1110 A.D. The code was based on the earlier codes and conventions current during the previous reigns. It is regarded as the first written constitution of Manipur. The later kings like Kyamba (1467-1508), Khagemba (1597-1652), Garibniwaz (1709-1748), Bhagyachandra (1763-1798) and Chourjit (1803-13) further expanded the decree.

The decree of Loiyumba, *Loiyumba Shilyen* deals with the distribution of occupations according to yumnak (lineage), assignment of duties to priest and priestess (Maibas and Maibis), assignment of the works of maintenance of the abode of deities (*Umanglais*) to selected *sageis*, creation of administrative departments (Loishangs), duties and functions of kings and queens, royal etiquette, titles and decorations awarded to the nobles, administration of justice, keeping of standard time and many other things (Khulem 1975). In the case of well-developed industries like handloom and dyeing, the weaving of different types of cloths was assigned to different yumnaks or sageis belonging to various *yek-salai* (clan).[8] Those occupational weaving crafts were distributed to 34 lineages. It is further reported that most of the traditional dyes were produced and developed under Loiyumba's patronage. The tasks of dyeing were distributed to eight lineages. It is mentioned in *Mashin* (manuscript dealing with the duties allotted to the people) that the color of the yarn used in weaving the *Ningthouphi* (King's cloth) was dyed in diverse colours by different lineages (Devi 1998: 87). The assigned occupations of the lineages were not interchangeable. Those who did not conform to the orders of the king were punished by being sent into exile or by reduction

[8] The main feature of the social structure of the Meiteis is the institution of *Yek-Salai*. It is a large exogamous unit, each tracing itself to a common mythical ancestor, who is part of Meitei Divine Pantheon. It can be loosely translated as clan. Meitei society has seven patrilineal units known as *Yek-Salais*. These seven *Yek-Salais* are *Mangang*, *Luwang*, *Khuman*, *Moirang*, *Angom*, *Khabanganba* and *Sarang-Leishangthem* (also called *Chenglei*).

to a degraded class or status of Loi. This system of strict ascribed specialisation is a vivid example of 'patriarchal feudalism'. It did not give the women weavers the freedom to engage in diverse weaving crafts.

Name of lineage	*Weaving craft*
Thingucham	Mungphi
Lairellakpam	Laiphi
Kaswam	Wana Katang Phi
Chakpa Thiyam	Laiphi Tonkap
Chakpa Lampam	Phiren
Kongpacham	Langtomphi
Thonngkabam	Yarongphi
Khumallabam	Chamiphi
Haokhom	Tungkap Phi
Laikhuram	Sarong Phi
Salchiram	Leirum Phi
Wahengbam	Dolai Kup Phi
Kabacham	Dolai Thanaba Phi
Apucham	Phiren Phi
Haodijam	Heikoi Phi
Thodingcham	Singkap Phi
Thongpam	Laiyek Phi
Kongbam	Purum Phi
Chingkhalpam	Tarao Phi
Thinpam	Pumthit Phanek
Taibangcham	Phimu Yekpa Phi
Yumlenpam	Nongphi
Wokthiyambam	Khunung Musum Phi
Chakpram	Phiren
Thoudam	Phagang Phi
Nongthombam	Ningthouphi
Ahongsangbam	Embroidery work
Ninghthoujam	Chinphi
Khoirisumbam	Khunung Chumkhun Phi
Yumnak Sangbam	Chingkhong Phi
Khoisnam	Phimu Lanphi
Singkhangbam	Khunung Mathang Phi
Chingkhwapam	Korou Phi Phingang
Wangpacham	Shangmi Lanmu Phi

The hierarchy in the social pyramid had to be expressed through the products of the female craftsperson. Consequently, instead of elaborate paraphernalia of regal grandeur, the colour, size, shape of the simple wearing apparels and ornaments were used to signify the status of the chief and other dignitaries. This became an important factor in the development of crafts work in the society. To differentiate the chief from the serf, the craftsmen (craftswomen in this context) had to produce goods of superior appearance and durability in terms of colour, design and quality. The craftswomen were always on the attempt to improve their skill and they had even to experiment with vegetable dyes and different threads viz., cotton, silk etc., to attain excellence (Haobam 1991: 207).

It is further reported that most of the traditional dyes were produced and developed under Loiyumba's patronage. The tasks of dyeing were distributed to 8 lineages as follows:

Name of lineage	*Colour of yarn*
Akangcham	Leirel Khoipaklei (light pink)
Chingkhwapam	Chengsang Muyumlei (not identified)
Yiroibam	Ureirom Makong Meiri (saffron dark flame red)
Khulmanthempam	Higok Nanou Lei (light blue)
Khumukcham	Kumlang Kumpal Lei (black)
Yipusangbam	Chingya Napu Lei Heitroi Lei (grey, yellow, green, deep maroon)
Washiyampam	Wana Katang Namba Sanglen Yaichu Lei (greenish red shade produced by mixing green and red threads)

It is mentioned in *Mashin* (manuscript dealing with the duties allotted to the people) that the color of the yarn used in weaving the *Ningthouphi* (King's cloth) was dyed in diverse colours by different lineages (Devi 1998: 87).

Colour of yarn	*Name of lineage*
Loiren Hoipak Lei (rose colour)	Akangjam
Mayumlei (light brown)	Chingakham
Ureirom Makhong Meiri (flame colour)	Irom
Loirang Thangchulei	Phijam
Higok Nanoubi (blue)	Khumanthem
Chingya Napurei (yellow)	Ipusangbam
Sanglen Yai Machu (deep green)	Washiyambam

It is interesting to note that there is also a dimension of sex role differentiation in weaving. Male weavers used to come to the weaving house of the palace if particular clothes had to be made for the Kings and Queens (Bahadur 1997:12). The weaver of *Ningthouphi* (royal apparel) had to be a male and was generally the chief of the weavers in the state. There seems to be a historical intervention when a male was made the head of all weavers of the land. This is in sharp contradiction to the mythological ontology discussed above. Women were excluded from weaving the royal apparel since they menstruate and therefore considered unclean. This whole arrangement raises the question of sex role differentiation in this field since weaving is generally an activity of womenfolk in Manipuri society. This also bears the imprint of the patriarchal social structure where the main weavers (female) are placed in relatively powerless status. During the days of the Maharaja under British domination, there were idiosyncratic rules but not oppression. For example, Maharani Dhanamanjuri Devi liked 'Thambal machu phanek' (pink lotus colored Manipuri sarong dress of ladies) and no one was permitted to wear it except the Maharani.

Another dimension that we can also examine here is silk manufacturing. The manufacture of silk cloths is entirely in the hands of the Lois, whose low social position permits them to practice many remunerative forms of employment, which customs deny to the Meiteis (Hodson 2003:27-28). The Loi being the cultivators of silk is fatal to their upward extension, as due to the custom of the country that associates position or caste with the nature of the various employments pursued, anyone wishing to engage in silk culture must lose position and will become a Loi. This is because after Hinduism had been adopted as the state religion since the early 18th century the issue of pollution and purity has become a major significance in the Manipuri society. As per the 'new Hindu social system' since silk culture involves killing of silk worms, it is considered as a polluted activity. Thus, the production of silk was on a very limited scale. However, today anyone either belonging to Loi community or not can engage in sericulture. This activity has become one of the main occupations and major source of income among the women in Manipur.

Gender construction of weaving

The dress and the mode of dressing forms a significant part of the Meitei tradition. The Meitei cultural tradition of the art of weaving is based on a mythical foundation. According to mythology, it was the supreme goddess of the Meitei traditional pantheon or the primeval celestial ancestress who introduced the art of weaving. It is believed that *Chitnu Tamitnu*, a goddess, discovered the cotton and then produced the yarn. When the threads are ready for weaving she arranged the required equipments and constructed the '*Sinnaishang*' (work shed). It is also believed that the goddess *Panthoibi* once saw a spider producing fine threads and making cobwebs and from it she derived the idea of weaving and thus started weaving. The goddess of weaving symbolizes wealth and prosperity. Again the feminity of the deity who introduced the art signify the association of this occupation with the females (Ibid: 106).

Weaving was so much culturally esteemed in the traditions of the Meitei that in the community ritual of Lai Haraoba, the dance performed by the Maibis symbolically taught the young girls the art of weaving. It was also, reportedly, a practice for the parents to give a set of weaving apparatus to their daughter as indispensable item of marriage gift on her marriage. It would be worthwhile to note that at the time of selecting a daughter-in-law, the Meitei parents gave priority to the prospective bride's excellence in the art of weaving. Until recently, it was compulsory in the Meitei society for every female to be trained in the art of weaving as it was held that females without the knowledge of the art of weaving did not possess the quality of womanhood, and for that matter, they had not the source of wealth and prosperity. That is why the knowledge and skill of weaving was compulsory for every Meitei female. Acquiring knowledge of the art of weaving was thus the greatest asset to a girl of marriageable age. The women provided their families all the necessary household cloths by making the cloths themselves. It is reported that every household had at least a set of weaving looms, either the loin loom (*Khwang Iyong*) or the throw-shuttle loom (*Pang Iyong*) type.

Unlike in other parts of India, weaving in Manipur is entirely the work of women. Weaving is part of their domestic duties. In fact, it is primary qualification of a Meitei girl. This handloom industry is practically monopolized by women not only from the idea of economic necessity but also from the sense of social custom. A Meitei girl is initiated in this art at a tender age, and throughout her life, she practices this art. Not only she supplies clothing to her family members but also makes it a source of family industry. It is said that the development of this industry by women is linked with the fact, that the Manipur men were continuously engaged in wars, and these women supplied them with the uniforms and maintained their families from the sale product of this industry (Roy 1979: 13).

The fact that various types of clothes were used is revealed in the writings of the hand written *Puyas* (ancient manuscripts), though no date has been mentioned. In addition, the oral traditions of the Meiteis such as legends, folk tales etc. clearly indicate that various items necessary for weaving of clothes were included as dowry[9] of the princesses and daughters of royal families. For example, at the time of daughter's marriage, different items of weaving such as 'Khwang Iyong' (loin loom), 'Pang Iyong' (throw shuttle loom) etc., were some of the essential items of dowry. There is also a reserved room known as 'Ningol Ka' (daughter's room) in the traditional Meitei house called 'Meitei Yumjao'. The unmarried girls of a particular locality would bring the items used in weaving like *Tareng* (spinning wheel) to assemble in this room and performed the work of spinning till late night and engaged themselves in a sort of competitions for bringing out the best product. This tradition was known as *Sinnaipham Kaba* (going to work). This was done on a rotation basis from one house to the other or every other day. This practice began to be discontinued after the World War II (*ibid:*9).

Handloom Industry Under Colonial Patronage

Commercialisation, Modernisation and Institutionalisation

When relations between Manipur and Britain got underway in 1763, the clear mention of giving silk to the East India Company by the king of Manipur, Gour Shai, shows that silk was widely produced and used in Manipur

---

9. Dowry here refers to a kind of gift to the bride at the time of marriage. It generally consists of items needed to start a family. The choice of items are decided solely by the bride's parents. However, there are certain items which should be included as per the custom of the society.

(Bahadur 1997:12). After the country completely got rid of the Burmese depredations in 1826, a more congenial political climate prevailed for the recovery of the craft. But due to the lack of communication and other facilities, the craft remained in a languishing condition until recently.[10]

Since the beginning of the second half of the 19th century the British became the additional partner of trade with Manipur. It was only in the later part of the nineteenth century that mill cloths of Great Britain began to be imported to Manipur. The importation of English yarns led to the decay of spinning in Manipur. But weaving continued to survive and the cloth made by Manipuri women with the silk yarn was getting market in other parts of India (Singh 1963: 81-82). There were some particular articles, which the Manipur Government strictly banned in exporting, and the one who violated this restriction was to face heavy punishment. The following articles were not allowed to export from Manipur, as they were meant as present, etc., for services rendered to the state:[11]

1. Khamen Chatpa — Coloured silk dhoti
2. Numthang Kulel — Laced puggree
3. Ningthou Phi — Shirt like garment
4. Lai Chuppa — Coloured shirt.

[10] Manipur's economic recovery after the end of the Burmese occupation in 1826 proceeded along with streamlining of the administrative structure and some liberalization of modus operandi of economic transactions as carried out under the British influence and even mild pressure at times. The British policy vis-à-vis Manipur was shaped by the objective of putting on Manipur as a strong defensive base against aggressive designs of Burma. They also sought to use the territory of Manipur to open up an overland trade route to Burma and there from to Yunnan Province of China. In pursuance of these objectives in nurturing closer political and military ties with the king of Manipur and making agreements with him to facilitate and promote flow of trade between British India and Manipur and Burma with Manipur's cooperation.

[11] *Annual Administrative Report of the Manipur Political Agency*. 1873-74. p. 14.

The annexation of Manipur in 1891 opened new vistas for the large-scale influx of finished products from Britain as well as from other parts of India. For the British, Manipur was not a potential market because of its underdeveloped or subsistence economy as well as the meagerness of the population. But the opening of Manipur to the outside world through Imphal-Dimapur road and Imphal-Jiri road led to the inflow of ever-increasing commercial items to the markets of Manipur. The shifting of administrative power from Raja to the Political Agent and inflow of such commercial goods brought a dramatic change to the traditional manufacturing industries. Some of them simply faced extinction due to lack of royal patronage, some survived the onslaught of capitalist goods, some other flourished with the application of modern technology and some completely new industries also developed at the same time (Singh 1998: 115).

From among the traditional manufacturing industries, weaving occupied an important place. It was this industry, which a successful modernization was done, and the products attracted the Europeans and Indians alike. By 1900, the valley people became adept at weaving in fly shuttle looms. In 1905 the chemical dye was available in the Imphal market while fast chemical dye appeared in Imphal in 1930. Various steps were taken by the administration to modernize the weaving since 1910. An expert weaver was employed for 3-4 months in this year. He instructed two Manipuri women in the use of improved handlooms of Salvation Army's Automatic Handloom.[12] Later on a weaving school was also established. The two women were appointed as instructresses in the weaving school.[13] As a result of the concerted effort of the administration and the weavers of Manipur, weaving became the most valuable cottage industry in the state by the

[12] File.No. 138. Office of the Political Agent in Manipur. Records of the Government of Manipur, Secretariat Library, Imphal.

[13] *Administrative Report of the Manipur Political Agency*, 1911-12. p. 7.

beginning of 1920s. During these years the export of Manipuri cloth rose considerably due to the Swadeshi Movement in India. Manipuri cloths were strong and durable. The better qualities were remarkable for their artistic design and elegant finish.[14]

In Manipur, weaving of cotton cloths was more popular than that of jute cloths. So, in 1921 there were 26923 weaving and cotton sizing workers against 10 jute spinning, pressing and weaving workers.[15] It was in the year 1921, Hairen Keifa Singh Selungba, an expert craftsman of Yaiskul Hiroohanba leikai, Imphal, introduced fly shuttle loom in Manipur. Earlier he had been to Tripura on a short visit and had picked up the new skill there. Although he did not succeed in his attempt to popularize fly shuttle loom, he made a number of Manipur young men conscious of the new possibilities of the craft and helped to widen their vision. In 1922, K. Tomba Singh of Yaiskul Hiroohanba Leikai and H. Amooba Singh of Uripok Haobam Dewan Leikai, Imphal left Manipur for obtaining regular training in the weaving industry in Serampore Training Institute, West Bengal. After two years, the craft received tremendous support from an unexpected source. In 1924, Mrs. Jolly, private tutress of the Maharani Dhanamanjuri Devi became interested in the handloom industry of Manipur. She engaged some expert Manipur female weavers and started producing various fabrics of European taste by using the traditional designs found among the Meiteis and the various tribal groups of Manipur. What is important is that their fabrics were produced with traditional looms like Pang Iyong (throw shuttle loom) and Khwang Iyong (loin loom). The team consisting of Mrs. Jolly, K. Tomba Singh and H. Amooba Singh did a great deal in reorganizing the handloom Industry of Manipur. They

---

[14] *Administrative Report of the Manipur Political Agency*, 1921-22. p. 11.
[15] *Census of India*. 1921. Vol. I. Part II, table-XVII.

collected design from the hills and valley of Manipur and successfully improvised them with the modern technique and introduced them in Manipur. Finding remarkable improvement in handloom industry of Manipur, Capt. Harvest, the then President of Manipur Darbar permitted to open a weaving institute of "Arts and Crafts" in 1930 and appointed Mrs. Jolly as the first Manager of the institute. However, on her leaving the state for good, His Highness, Maharaja Churachand Singh appointed K. Tomba Singh as the Manager of this Institute in 1936 (Roy 1979:12). This Arts and Crafts institute/department soon flourished and during the year 1932-33 the total expenditure on the department was Rs. 14,827 and the total receipt was Rs. 12,233. There was also stock in hand in the value of Rs. 9,824 and collected bills amounting to Rs. 1,500 approximately. These figures show that the department was a profitable one.[16]

By establishing 'Manipur Handloom Industry' in 1936, attempts were made to export handloom products outside Manipur and India. In 1939, main items like bed-sheet cover, cushion cover, tray and coy, breakfast table napkin, dressing gown, dinner table mat etc. were produced and exported not only to Indian states but also to Europe, Australia and Africa and these were purchased and liked very much (Bahadur 1997:12). After the bombing incident at Imphal during the World War II in 1942, the building of weaving institute Arts and Crafts and along with records, equipment and everything in the building was destroyed. But after the war in 1945 the department was reorganized and the office was used in a room of the Land Revenue office building.[17] Soon this weaving industry recovered from war ravages and the Arts and Crafts department added to the state exchequer an amount of Rs. 1,17170 in 1946.[18] So, this industry flourished vigorously

[16] *Proceedings of the Manipur State Darbar*. R. N. 1 of 9.8.1933, Imphal.
[17] *Administrative Report of the Manipur Political Agency*, 1944-45. p. 7.
[18] *Administrative Report of the Manipur Political Agency*, 1945-46. p. 11.

under the patronage of the colonial rule. With the merger of Manipur in the Indian Union on 15 October, 1949 there have been a lot of measures taken up for the development of handloom industry.

Handloom Industry Under State Patronage

'Cooperatisation', entrepreneurship and women empowerment

Around 1950, the staple yarn was imported into Manipur. During 1960-70, four-ply acrylic wool was brought in. This four-ply wool was split laboriously into one-ply or two-ply wool for weaving. Then, around 1980, one and two-ply wools were introduced and the hill people also began to use these wool yarns to weave cloths conveniently in assorted colours (Bahadur 2000:9-11).

Manipuri society has an inbuilt system of traditional socio-economic cooperation known as *Marup*,[19] which continues even today. However, the modern day cooperative movement started after 1947 when the Indian Cooperative Societies Act, 1912 had been adopted as the Manipur Cooperative Societies Act, 1947. As per the Joint Census of Handlooms and Powerlooms 1995-96, Manipur has the third highest concentration of handloom households with 8.55 per cent of the share of all India after Assam (43.38 per cent) and West Bengal (8.69 per cent). Moreover, Manipur has also got the second highest concentration of weavers with 12.26 per cent of all India share after Assam (36.30 per cent). There were 3.39 lakh weavers with 2.77 lakh different looms producing

[19] In Manipur, *marup* system is one of the most important and effective forms of traditional socio-economic organization for mutual help, including reciprocal gift giving, exchange labour pattern, assistance in death, illness, marriage and other personal crisis. It is, thus, much more than a simple modern economic organization. It provides both economic and social security of an individual in every sphere of social and economic life. It not only provides economic development in the region but also act as a means for strengthening the overall solidarity of the society.

86.40 lakh linear metre of handloom cloth per month. Out of the total weavers 70 per cent were working full time in the valley areas as against 40 per cent in the hill areas.

With the organisation of weavers' cooperatives in the state in the year 1950-51, a new era for handloom industry began. The weaving cooperatives were organized and established with a view to help the weavers particularly in the supply of raw materials, such as yarn and marketing of the products. The Manipur State Handloom Weaver's Cooperative Society was established in the year 1955 as an apex institution to play a pivotal role to procure and supply the required raw materials and equipments to the weavers and to market finished products of the societies affiliated to the State Society. There are now 1390 Handloom Weavers Cooperative Societies with 1,76,152 members (Government of Manipur 2005). In Manipur, it is observed that 29.59 per cent of the total cooperatives and 35.58 per cent of the total cooperative membership belong to handloom cooperatives. Out of the 3.39 lakh weavers in Manipur, 25.79 per cent were members of handloom cooperatives. According to NCUI Annual Report 1999-2000, 6.53 per cent of the total number of handloom cooperatives and 7.46 per cent of the members in India belong to Manipur. It is in this context that we are going to examine whether cooperatisation of handloom sector has empowered the weavers who are exclusively women.

A Case Study of Imphal East District

The total number of handloom cooperatives in Imphal East district is 330 constituting 23.74 per cent of the total 1390 handloom cooperatives in Manipur. It comprises 19.51 per cent of the total membership.

Out of the 330 handloom cooperatives, only four do not have the registration number as per the record of the Cooperative Department, Imphal East district. It was observed that the first handloom cooperative to be registered was in the year 1955-56. In the year 1988-89 the highest numbers of cooperatives were registered i.e., 40 followed by 36 during 1984-85. The years with not even a single registration were 1956-57, 1957-58, 1959-60, 1961-62, 1963-64, 1964-65, 1965-66, 1966-67, 1993-94 and 1998-99.

The present study tries to investigate whether the introduction of cooperatives in handloom industry in Manipur has really empowered the cooperative weavers who are entirely women. The research, which was conducted during August-October 2006, and August-December 2007, has covered five villages of Imphal East district of Manipur viz., Ucheckon, Bashikhong, Khurai Sajor Leikai, Khurai Thangjam Leikai and Khurai Khonangkhong. 35 handloom weavers' cooperative societies were selected for the study. Then, 250 handloom weavers belonging to these thirty-five cooperatives were chosen through random sampling. The primary data were collected with the help of specially prepared interview schedule. The technique of observation was also employed in conducting this research.

In the course of our investigation, we have found various constraining factors in the role of cooperatives as an agent of women empowerment. Most of the weavers work as part time weavers with less professional outlook. This is because of the existing social order where a housewife practically manages all activities in the family, which becomes a constraint on the production trend of the societies particularly in this age of rising commercial competition from mill-made clothes. The low average earning of a member per month may also explain the tendency for part time weaving. There also occurs the flexi-time theory in weaving sector of Manipur. Most of the part time weavers are involved in occupational multiplicity without much importance being given to professional competence on large scale. This has also resulted in the limited scale of production with static design and innovation.

There is lack of sound commercial organisation of the society. It has been observed that the societies do not function as commercial enterprises based on professional norms, financial disciplines and written bye-laws. From time to time, one could also find the problem of vested interest of a handful few who tend to dominate the process of decision-making, operational plan and implementation. The broad based discussion among the members with reference to competitive market is fairly low. There are also many one-woman societies where prolonged continuation of the same person as the sole executive is observed.

Another interesting observation is the use of land, either by the President or the Secretary for the construction of office building and workshed of the societies. The private land of the President or the Secretary is hypothecated in the name of the society on paper. Now the land legally belongs to the society. Office building and workshed are constructed on the land. The members of the family continue to avail of the benefit of accommodation for various purposes in the building being constructed. Cosy drawing rooms, kitchens and bedrooms are part of the building's premises. As such the purpose of the workshed and office becomes secondary. There is also a feeling among the functionaries of the cooperative societies for having the office within their easy reach.

It is equally interesting to have reference to a list of members of the weaving cooperative societies belonging to the same family and close relatives. The promoter of the society particularly presidents and secretaries have a tendency of getting the expected benefits monopolized within the limited framework of family members and relatives. In some cases, the societies have become a new form of individual property and a mere family enterprise. Thus, the consideration of competitive production becomes secondary while the primary consideration is to get full advantage of loans, grants and subsidies.

The weaving cooperative societies in Manipur have been experiencing the problem of fluctuating supply of yarn. The yarns of various kinds required by the weaving cooperatives come from different states like West Bengal, Assam, Tamil Nadu, Maharashtra etc. A few businessmen control the yarn supply of the Imphal market. This has created one of the serious constraints on the diversification of production. While some of the weavers have commendable skills of diversified production of competitive designs, production remains impossible with limited availability of yarns.

The study reveals that there has not been any change in the type of technology used for production in weaving industry in Manipur. The weavers still use the traditional loin looms, throw-shuttle looms and fly-shuttle looms for their production. This has increased the burden on the workforce and also increases the cost of production in the long run. It has also been observed that practically the cooperative societies are running mainly for the purpose of production with negligible attention to the marketing system. There are no marketing cooperatives based on set principles of large-scale marketing. This has been one of the major hurdles on the expansion or diversification of production of weavers' cooperative societies in Manipur.

The investigation reveals that there has been very low per centage of cooperatives and also members who received credit or financial assistance from the government. This shows that the societies in Manipur cannot avail of the benefit of economies of scale because of lack of sufficient credit support. Another factor, which affects the proper functioning of cooperatives, is the inefficiency and unaccountability of the government departments. The economic atmosphere has been largely disturbed by the unabated practice of corruption and equally affected by bribery, nepotism and favouritism.

Conclusion

From the above discussion we can derive a framework of evolution of weaving in Manipur from mythological, royal, and colonial to state patronage. Under the royal patronage the craft of weaving was lineage based and feudalistic. While there was modernisation and institutionalisation of the handloom industry in Manipur during the colonial rule, the processes of 'cooperatisation' and entrepreneurship were started under the modern democratic state. The framework can be represented as follows:

Under Royal patronage → Lineage based, feudalistic

↓

Under Colonial rule → Modernisation, institutionalisation

↓

Under the State → 'Cooperatisation', entrepreneurship

The question of women empowerment through handloom cooperatives is also investigated. A cooperative society is defined by the International Cooperative Alliance (ICA) in its Congress held in Manchester in 1995 as "an autonomous association of persons united voluntarily to meet their common economic, social and cultural needs and aspirations through a jointly owned and democratically controlled enterprise". The following seven principles were formulated in the Congress– i) Voluntary and Open Membership; ii) Democratic Member Control; iii) Member Economic Participation; iv) Autonomy and Independence; v) Education, Training and Information; vi) Cooperation among

Cooperatives; vii) Concern for Community. The cooperative form of organisation is solely designed for promoting the mutual interests of user patrons on the basis of equality and equity. They control is on a democratic basis. It is a local organisation, and provides for local participation. It is responsive to local needs, as its policy is decided democratically by the local member-users. It serves as a training ground for rural people in business and in democracy. Cooperatives are considered as significant tools to adopt participatory approach not only for the rural development but also for economic empowerment of women. Cooperatives not only provide economic empowerment but also socio-political empowerment to the people involved. A distinctive feature of these bodies is that they are organized on the basis of certain principles, the most important of which, from the sociologist's point of view, are equality among members and the democratic principle in management.

The study has revealed that there are many constraining factors in the role of cooperatives as an agent of women empowerment relating to management, production, technology, raw materials, marketing, credit, employment, and inefficiency and unaccountability of the government departments. Thus, we see that the basic ideology and principles of cooperatives are generally missing in the case of handloom weavers' cooperatives of Manipur.

However, it is worth noting that since most of the weavers belong to family below poverty line they do not have enough capital to invest in their businesses. Therefore, rural institutes like cooperatives are their only hope as one of the instruments for financial assistance, marketing and other benefits. In order to bring cooperatives as an instrument of women empowerment there is an urgent need for revitalization and re-organisation of handloom weavers' cooperative societies. The level of members' participation needs to be improved by sensitizing and motivating them before enrolment as members and guiding them to develop a professional outlook in their activities. This should be accompanied by efficiency and accountability from the concerned government agencies. Moreover, the relationship between state and cooperatives should be based on complementary model i.e., the state and cooperatives develop a symbiotic and mutually complementary relationship rather than administrative model in which the state has a domineering role in the cooperative development.

References

Bahadur, Mutua (1997). *Traditional Textiles of Manipur*. Mutua Museum, Imphal.

Bahadur, Mutua (2000). *Tribal Hand Woven Fabrics of Manipur*. Mutua Museum, Imphal.

Brown, N. (1975). *Statistical Account of the Native State of Manipur and the Hill Territory under its Rule*. Sanskaran Prakashak, Delhi.

Devi, K. Sobita (1998). *Traditional Dress of the Meiteis*. Bhubon Publishing House, Imphal.

Haobam, Shyamsunder (1991). 'Arts and Crafts of Manipur'. in Sanajaoba N. (ed) *Manipur: Past and Present*. Mittal Publications, New Delhi.

Hodson, T. C. (2003). *The Meitheis*. Akansha Publishing House, Delhi.

Roy, Nilima. (1979). *Art of Manipur*. Agam Kala Prakashan, Delhi.

Singh, L. Ibungohal. (1963). *Introduction to Manipur*, Imphal.

Singh, N. Basanta. (1988). *Socio-Economic Change in Manipur: 1891-1947*. Imphal.

13

Traditional Healing Practices in Darjeeling Himalayan Region

Santosh Kumar Rai

Himalayas are regarded as a great repository of natural wealth and biological diversity. The physiographic, climatic and edaphic conditions often aided by the biotic factors make the Himalayas an ideal habitat for a wide range of plants and animals including human being. In addition to their pristine beauty, Himalayas are an important watershed and play a vital role determining climate and vegetation for economic well being of more than a billion populations, more than any mountain system of the world.

The Indian subcontinent harbours diverse vegetal wealth and multiethnic communities where more than 7500 plant species and many of them of Himalayan origin are being used as medicine for the treatments of various ailments and diseases. The earliest records of use of medicinal plants in prevention and cure of diseases in India are found in the Vedic literature like *Rigveda* (ca 3500-1800 BC) and *Atharvaveda* (1500 BC) and primary surgical practices are also mentioned in *Charak Samhita* (1000-800 BC) and *Sushruta Samhita* (800-700 BC). The *Ayurveda* and *Siddha* systems of medicine also originated from the traditional medicine system of India (Jain 2000). The folk medicines obtained from natural sources are still in use in the remote areas of many under developed or developing countries that have no modern health facilities.

Darjeeling Himalayan region is one of important parts of the Eastern Himalayas. Sandwiched between Bhutan and Nepal from the east and west, and Sikkim and the plains of West Bengal from the north and south, a large fraction of population of this region lives in the villages and far-flung areas and is dependent on forest resources to meet most of their daily needs, including medicine, food, fuel, household articles and the materials needed for performance of rituals (Rai & Bhujel 2007).

People and Their Belief

Major communities in the region are the Bhutia, the Lepcha and the Gorkha, living here since time immemorial. They have their own traditional system of healing passed on from generation to generation. The existence of traditional knowledge for the treatment of different ailments and diseases with the help of medicinal plants is common among the spiritual healers of all the communities. Consulting the community priest is an indispensable step, before the treatment of any ailment among all the communities, particularly in the village. Even in urban areas, they usually give their first preference to the traditional system of treatment and choose alternatives like the modern system of treatment, later, if they have to.

Healing Practices in Bhutia

The Bhutia people are Tibetan descendents, original inhabitants of *Bhot* in the present day Tibet Autonomous

Region of China. They are followers of Buddhism. Their priest or spiritual healer, known as the *Lama,* conducts all the rites associated with their cultural and religious affairs like raising prayer flags. Their qualified and experienced medical practitioner is known as *Amji* and *Pow.* Medicines are given in the form of powder and pills obtained from different plants. In some cases gold, pearls and other mineral particles are also added for the synergistic action of the medical plants. During the course of treatment pulse rate of the patient is generally observed and diagnosis is made. Other factors like the patient's personality, astrological chart (both planetary and constellation based), age, climatic condition, diet, behaviour and the physical and spiritual surroundings are also considered. Healing is done through the correction of physical imbalances, though, nowadays the doctors rely more on the allopathic diagnosis of the patient through pathological laboratory reports (Rai and Sharma 1994). There are a few clinics of the Tibetan Medical and Astrological Institute (TMAI) in the district where the practitioners usually bring the medicines from Dharmasala (Himachal Pradesh). There is another system of herbal treatment with a deep root in the religion (Buddhism) and is spirituality known as *Yuthog Nyingthig.* It was the wish of the 5th Dalai Lama because of which Desi Sangye Gyatso established the Chagpori Medical Institute in Lasha in the year 1696. After the Chinese occupation of Tibet in the year 1959, the Tibetans engineered its rebirth in India (Dharmasala, Himachal Pradesh). Dr. Samphel Norbhu Trogawa Rinpoche brought it to the district of Darjeeling in the year 1992. The Chagpori Tibetan Medical Institute (CTMI) is located at Singamari (near Darjeeling town) with branches at Kalimpong and Takdah. The teachings in medicine, pharmacology, botany, *Dharma,* astrology, Tibetan language, Tibetan literature and English are carried on at this institution. A curriculum of five years theory and two years practice is the standard (Rai and Sharma1994). The CTMI team collects most of the herbs from Nepal, Sikkim and Bhutan and now they aim to start a plantation of important herbs within the district.

Healing Practices in Lepcha

The Lepcha people call themselves the *Rong* and locate their original home in the neighbourhood of the great mountain Kanchandzonga. They are of mongoloid descent and the place they live is called *Mayal Lyang*.[1] Their spiritual priest is called *Boongthing* for male and *Mun* for female. They are regarded as qualified and experienced traditional healers in the region. Consultation of the community priest is indispensable before treatment. The direct use of plant parts is found in most of the medical prescriptions, in the form of paste and water extracts. In some cases, animal parts particularly those of bears, deer and tigers are also added in their formulations.

Healing Practices in Gorkha

They are the dominant race in Darjeeling Himalayan region, basically cultivators and are resourceful and hard working. They taught agriculture to the aboriginal Lepchas (O'Malley 1907). Gorkha is a generic category that includes castes like Bahun (Brahmin), Bhujel, Chhetri, Damai, Gurung, Kami, Manger (Thapa), Newar (Pradhan), Rai (Khambu-Kirat) Sarki, Sherpa, Subba (Limbu-Kirat), Sunuwar (Mukhia), Tamang (Murmi), Thakuri, Thami, Yakha(Dewan) and Yolmu(Kagate) with much diversity in their religious faith and languages (Subba 1992). With the exception of a few, most castes are animist and shamanism is associated with their caste and clan, culture and tradition recognized by the society. Medicines prepared from plants are known as *jaributi* or *pahare dabai* and exist in the

---

[1] The Lepchas, according to the mythology, are Rongpas (ravine dwellers) and their tribal homeland is referred as Mayal Lyang ('*Hidden Land*' or '*the land blessed by God*'). They are believed to be the aboriginal inhabitants of Sikkim, India, known for their retention of rich cultural heritage.

practices of shamans known as *Dhami* and *Jhankri* and the enthusiastic village folks known as *Baidhya*.

The direct usage of plant parts with different ingredients is found in most of the medical prescriptions, in the form of paste, decoction, water extract and pills or tablets. In this system also some of the ingredients are animal parts, like in the Lepcha system. This system has not attained the status of the Tibetan herbal system. However, a majority of the population in the region follows this traditional system. In this context, family members of late Baidhya Chewang Pakhrin are worth mentioning with their successful practices in herbal drugs. They have been engaged in treating bone fractures with wild herbal medicines over the last 55 years. They claim that this knowledge is inherited from their forefathers and is successfully being practiced by the fifth generation now. The secret knowledge of healing is said to have come to the first Baidhya (Vaid) of the family as a blessing in his dreams. The Late Baidhya Chewang Pakhrin built an institute with a 30-bed hospital constructed with the help of donations and public contributions. The institute was inaugurated on 27th February 1995. The institute is located at 16th mile, Kalimpong in the district of Darjeeling and is of immense benefit to the people of Darjeeling and its adjoining regions. The Institute is affiliated to the Council of Alternative Medicine, Calcutta (Bhujel 1996).

The Ailments and their Treatment

Treatment of ordinary disease and ailments is known by almost all people in the remote villages and far-flung areas. In case of medicinal plants, the herbal healer or *Vaidhyas* believe that all plants have a limited active principle for effectiveness. It is, however, difficult to ascertain which period or stage of the plant is more effective against the disease (Purohit *et al.* 1986). In some cases the spiritual healers practicing herbal remedies have made it mandatory to strictly follow sociocultural taboos. They have a particular time (early morning without meeting any person or sometimes evening), particular day (Tuesday, Thursday and Saturday), period (full moon) and the direction or position of the plant parts (east facing, south facing) etc. for the collection of medicinal plants. Addition and deletion of ingredients, concentration of the drug and doses are made according to the type of ailments and condition of the patients. Most of them maintain relative secrecy about the plant parts and ingredients used for their formulations. *They believe that the medicine will lose its healing power if too many people know about it.* They might even meet with some ill fate because of the wrath of the presiding deity of the medicinal plants in their forests falling on their heads (Biswas & Chopra 1956). Most of the diseases or ailments to be treated are burns, cuts wounds, sores, itching, skin irritation, cell inflammation, bone fracture, joint dislocation, sprain, rheumatism, fever, body ache, cough, common cold, influenza, flu, sinusitis, eye infection, tooth ache, mouth sores, tonsillitis, pneumonia, abdominal pain or stomach colic, gastritis, food poisoning, cholera, diarrhoea, dysentery, liver cirrhosis, ulcer, piles, urinary troubles, diabetes and heart troubles, and some of the treatments indicated are anthelminthic and wormifuge after proper diagnosis. The treatments of dog bite, snake bite and scorpion sting require really skilled hands and practices. In case of food poisoning, the diagnosis of the root cause is very difficult and only the local priest or shaman can diagnose and give proper treatment. The ailments of liver and stomach, locally known as *nash, kapat* and *harital,* are believed to be intentionally caused by the 'evil eye' as a punitive action or as a part of revenge or a part of jealousy by sorcerers, or sometimes the spiritual healers themselves. Sometimes in such cases, the victim may die of improper detection of causal agents or improper treatment. Only the experienced shamans can release the patients from trouble.

Shamanism and the People

Except a few ethnic group of religious faith, most people attribute illness and misfortune to a variety of supernatural forces such as attacks by witches, sorcerers, forest-divinities, spirits of deceased individuals and angry gods and goddesses. It is recognized that the human neglect is usually the basis for supernatural discontent, which invites attacks by witches and evil spirits. Through human failure and neglect supernatural forces are unleashed bringing chaos, disorder and disaster. These forces are combated, held in check or defeated by individuals who are capable of 'seeing' and 'identifying' them. Such individuals receive their power primarily through successful encounter with supernatural forces in the past (Hitchcock and Jones 1976). These individuals or shamans are known as *Jhankri* (in general), *Bijuwa, Mangpa* in Rai community, *Boongthing* (male) and *Mun* (female) in Lepcha community, *Phedangma, Yeba* (male) and *Yema* (female) in Limbu community, *Puimbo* (male) and *Ngiami* (female) in Sunuwar community, *Bonbo* in Tamang community, *Pajyu, Khepre* or *Ghyapring* in Gurung community, and *Lama* (male) and *Doloma* (female) in Bhutia, Sherpa and Yolmu communities. It is believed that whenever necessary these individuals can introduce spirits voluntarily into their body. Now, however, only the Rai and Limbu communities have deep faith in shamanism and worship their departed ancestors, nature, different deities and ghosts. They believe that their life is controlled by different deities. They use the service of *Bijuwa, Mangpa* and *Phedangma* who are believed to be capable of taking the soul of a dead person of their community to heaven by performing *chinta* (séance) at the time of last rites. They go into a trance while reciting the religious verses known as *mundhum* during the journey to heaven and back to earth. They wear a white frock or gown, headgear with feathers, *rudrakse* (a holy rosary) interwoven with *cowree* (seashell) and tiny bells across the body. They have the sociocultural rights to perform all religious rites at all places, according to caste and clans. Wearing this dress is compulsory during all of the rites and ceremonies performed by them. They are, however, supposed to qualify to wear it before their teacher in the beginning. Other shamans hardly wear any specific dress unless it is necessary, except a turban or a traditional cap, which is a requisite while performing rites and ceremonies. Another name given to *Jhankri* is *Dhami* (and sometimes *Baidang*). It is believed that the *Baidangs* do not have any spiritual knowledge of divine gift but they do know some incantations (mantras), as learnt from their teachers and practice the same at the time of need. *Vaidhyas,* also, do not have any spiritual knowledge but they have sound knowledge of herbal formulation. With the progress of the modern technology and medical science, the concept of a shaman is gradually vanishing. *The modern society has branded shamanism as a superstition.* But it is a segment of the Gorkha tradition, full of unsolved mystery, and needs proper investigative study (emphasis added) (Liwang 2000).

Shamanism and the treatment

For diagnosis, the shamans have a plate (usually made of bronze) for *jokhana*. In the plate, a pinch of unbroken rice grain is placed in three groups and counted in pairs. After a long run of counting of even or odd numbers they predict the omen, good or bad, responsible for the ailment.

In case of indigenously prepared medicines, they utter incantations meant to activate the healing power of the formulation before application and administration. They also claim to be capable of accelerating and reducing potentials of the drug or medicine in the body of patients after administration. Many old aged individuals in remote and far-flung villages have never taken any allopathic medical treatment in their entire lives or consulted a qualified doctor.

But their succeeding generation, presently in their middle age has a partial inclination while the next generation shows preference for modern medical facilities. One of the reasons for this change, according to them, is the non-availability of appropriate medicinal plant parts for the drug preparation while the other reasons include endorsement of the modern society rather than a cure for their ailments.

Conclusion

The prevalence of traditional healing practices among ethnic communities is always noted in places away from the modern world of amenities and comfort. The greater dependence of people on traditional system of medical treatment has a direct association with their habitats, which are situated among difficult natural barriers and constraints. The lack of modern hospital facilities and support from the government and the so-called civilized society has also forced them to follow their indigenous practices. The different ethnic communities living in the Himalayan regions hold vast knowledge of plant uses for different purposes including health care. Many of the medicinal plants and healing practices adopted by the shamans of different communities should be investigated scientifically so that their desirable properties can be useful for the welfare of large groups of people.

References

Bhujel, R.B(1996). A successful herbal drug practitioner of Darjeeling Hills: Baidhya Chewang Pakhrin—A profile. *Proc.Medicinal plants, herbal drugs and rural health*. Max Muller Bhavan, Calcutta. pp.86-88.

Biswas, K.P. and Chopra, R.N. (1956). *Common medicinal plants of Darjeeling and Sikkim Himalayas*, Alipur, Calcutta.

Foning, A.R. (1987). *Lepcha: My vanishing tribe*. Sterling Publications, New Delhi.

Hitchcock. John T. and Jones Rex L. (1976). *Spirit Possession in the Nepal Himalayas*. Vikash Publishing House Pvt. Ltd., New Delhi.

Jain, S.K. (2000). Global resurgence of ethnomedicobotany- The Indian scene. *Journ.Trop.Plants*. Vol.1(December):75-81.Malaysia.

Liwang, Min(2000). *The Gorkhas-a glance at their culture and traditions*. Himal (India), Kalimpong Publication, Kalimpong, DGHC.

Miller, Casper J. (1979). *Faith healers in the Himalayas* (Revised edition 1997). Book Faith India, New Delhi.

Morris, C.J. (1933). *The Gurkhas- an ethnology*. B.R Publishing Corporation. Delhi.

O'Malley, LSS (1907). *Darjeeling district gazetter*. Gyan Publishing House, New Delhi.

Pradhan, Kumar (1991). *The Gorkha Conquests*. Oxford University Press. 5 Lala Lajpat Rai Sarani, Calcutta.

Purohit ,V.P, Silas, R.A and Gaur, R.D. (1986).Vaidyas and herbal remedies in north Bengal. *Folklore*.27 (8):155-161.

Rai, Lalit Kumar and Sharma Eklabya(1994). *Medicinal plants of the Sikkim Himalaya: status, usage and potential*. Bishen Singh Mahendrapal Singh, Dehra-Dun.

Rai, Santosh Kumar (2002) Studies on the Ethnobotany of Darjeeling Himalaya. PhD thesis submitted to the University of North Bengal, Raja Rammohunpur, Siliguri-734430 (Unpublished).

Rai, S.K., Bhujel, R.B. and Das, A.P. (2007). Ethnobotanical studies in Darjeeling Himalaya in relation to birth, marriage and death. *Proceeding of seminar on Advances on Ethnobotany of the Indian Association of Angiosperm Taxonomy*, T.M.Bhagalpur University, Bhagalpur.

Rai, S.K. & Bhujel, R.B. (1999). Some less known Ethnomedicinal plants of Darjeeling Himalaya. *Jorn. Hill Res.* Vol.12(2):160-163. Gangtok, India.

Risley, H.H (1891). *Tribes and Caste of Bengal. Ethnographic Glossary* Vol. 2. Calcutta, Firma, K L, Mukhopadhay, Reprinted in 1981.

Singh, K.S. (Ed.1993). *People of India-Sikkim* Vol.-XXXIX. Anthropological Survey of India, Seagull Books, Calcutta.

Subba, Tanka B. (1992). *Ethnicity, state and development - a case study of Gorkhaland movement in Darjeeling*. Vikash Publishing House Pvt. Ltd. New Delhi.

Tsarong, T.J. (1986). *Handbook of traditional Tibetan drugs.* Tibetan Medical Publications, Kalimpong, Darjeeling.

14

Present Status, Scope and Future Prospects of Floriculture in Uttarakhand

Deepti Singh

Floriculture is a rapidly growing sector of India that deals with the production, utilization and marketing of ornamental plants and flowers. Floriculture is a totally skill-based farming, where everything from the seed sowing to packaging, is done by farmers. The world trade data shows that the share of developing countries amounts to more than 20% of the global market, while the Indian contribution in floriculture has not been more than 0.6%. In India, the total area under floriculture is 1,67,000 ha with the production of 987 MT loose flowers and 113618 million cut flowers (NHB, 2010). The average productivity of flower crops is quite high in comparison to other field crops.

In India, Uttarakhand with its different agro-climatic zones is particularly conducive to commercial horticulture and floriculture. The high altitude, cold winds and moist soil of the region make it ideal for floriculture, which requires low temperatures. Floriculture here is, thus, being developed in a big way in order to meet the demand of both domestic as well as foreign markets. The climate is ideal for growing flowers all year round (Misra *et al.*, 2008). Following the footsteps of the leading state in floriculture, Uttarakhand is making steady progress (Tyagi, 2011). The area under the state for flower production has almost doubled from 512 ha in 2005-06 to a 1000 ha in 2010-11 (Singh, 2011) and a little progress has been made on the export front. Hundreds of villagers in Uttarakhand are increasingly taking up floriculture as a result of the growing demand of flowers across the globe. More and more farmers are giving up traditional farming for floriculture in the state. The farmers produce exquisite commercial varieties of flowers like carnations, lilies, chrysanthemums, gladioli, gerbera and Indian red roses that are in high demand in domestic as well as in European markets. Uttarakhand has now emerged as the largest producer of Gerbera in the entire region of northern India. Lately, Marigold cultivation is up in a big way in Rudra Prayag area as a commercial crop for Char Dham Yatra. At present, the total turnover of cut flowers from Uttarakhand is over 55 Crores. In Uttarakhand, places like Padampuri (Nainital), parts of Dehradun, Udham Singh Nagar, Ram Nagar and Haridwar are now recognized as floricultural clusters (Singh, 2011).

Realizing the scope of earning more foreign exchange through the sector, the Uttarakhand state government declared floriculture as a high thrust area and announced several policies in favour of the production of high quality flowers. This led to large investments in this area and as a result, a large number of floriculture units, based on green house technologies for the export of their produce, were set up. These developments had an effect on the flower cultivation in the open field condition and thus, significant growth in the floriculture sector took place.

The entrepreneur (the farmer) has been found to play the crucial role in the innovation process and in the effort to harness the benefits of the quantum jump created in the market by the consumers. About two decades back or so, the floriculture was just a pastime of the rich and a hobby for flower lovers, but now it has opened a new vista in agri-business. For boosting the floriculture business, the state government has planned to rope in private companies to give fresh impetus to the production of various flowers. Advanced training centers in IHITC would enable incorporated learning and training to create strong market intelligence (Ladha, 2011).

With the spread of commercial floriculture activity in India, many nurseries have come up in the market with planting material for different floriculture crops. But the available planting material is not genuine, is infected and is of inferior quality (Arora, 2001). Apart from this, most small growers due to their scale of operations do not have access to the new improved varieties, which were imported to the country in recent years. Also, in spite of Uttarakhand making big strides in floriculture, it still has a long way to go to become a major floriculture hub that caters to the global market. This is a paradox since the state, nestling on the lap of the Himalayas, is endowed with rich natural resources all year round. The Government of Uttarakhand through its nodal agency State Industrial Development Corporation of Uttaranchal Ltd (SIDCUL) is establishing a floriculture project enabling comprehensive project execution inclusive of machinery supply, construction, management and operation of the entire setup to promote the floriculture sector by growing high quality bulbous material. SIDCUL aims to promote and scale up the development of the floriculture industry in the state, with the participation of local hill farmers and entrepreneurs by use of the latest technological intervention, and as a means for creation of employment and income generation opportunities for farmers in the state. Also, the state government has signed an MOU with Indo Dutch

Technologies Pvt. Ltd for the development of special infrastructure facilities and floriculture planting material. This would foster the promotion of export quality flowers from the state.

The Uttarakhand Government in its ambitious project of developing floriculture sector as a thrust area envisages an investment of Rs. 200 crore in this sector in the 11th plan period. After being carved out of Uttar Pradesh, the state has made significant progress in floriculture with government unabatedly showing special interest in developing floriculture as an industry. The focus now is both on quantity and quality with a view to make an impression on the global market. Floriculture in Uttarakhand has the potential to specialize in cut flowers, planting material like flower bulbs and seeds, and essential floral oils. There are more than 300 kinds of flowers and about 50 kinds of greens, which can fetch a good price in the Netherlands and Japan markets (Bhattacharjee, 2006). In order to capitalize on the climatic diversity and geographical features in Uttarakhand, the state plans to formulate measures for facilitating the commercialized cultivation of more than 300 varieties of flowers.

Scope

The scope of floriculture in Uttarakhand has increased tremendously. There is high competition as floricultural economy has shifted to consumer-driven enterprises, rather than a producer-driven one. In this scenario, producers of all sizes have to focus on marketing with a lookout for national brands to boost their sales. Growers have to look for niches and value-added products. There is a trend for direct marketing besides marketing, through super markets and wholesale markets. Business management is becoming important in this scenario of global competition (Chadda and Chowdhary, 2007). The domestic consumption of loose flowers, especially of Marigold, China Aster, Jasmine,

Crossandra and Barleria has been increased tremendously. Research is required on developing high yielding varieties and year round production of Chrysanthemum, China Aster, Marigold and many potential flowers in promotion of crops like Annual Chrysanthemum and *Desi* Rose. Floriculture is not limited only to commercial production of cut flowers and ornamental plant but also creates scope for subsidiary agro-business like export-import, nursery, seed production and organic manure production, and agro-industries like manufacturing of pigments, horticultural implements, drip and sprinkler irrigation systems, bio-fertilizers, tissue culture, modern packaging, extraction of essential oils and performing other value addition work. All these in return, create a huge potential for sustainable employment generation, both direct and indirect (Desh Raj, 2005).

The establishment of model floricultural nurseries is very important for timely supply of quality planting material. Co-operative florist organizations like Cut Flower Growers' Association exist but there is also a need to establish co-operative societies for traditional flower growers. This would help establish a minimum support price. Effective planning and monitoring is also very important from a quality point of view. Beside this, proper transfer of technology is also the need of hour.

It is pertinent to not that floriculture is emerging as a blooming business today. The jobs related to this field comprise handling plantations, nurseries and other landscaping projects. These works need scientific management of soil preparation, sowing, harvesting, administration of nutrients and controlling the climate. One can work in different capacities such as consultants, farm/estate managers, plantation experts and plantation supervisors, project coordinator. Research and teaching are some other opportunities. Marketing of Floriculture products for different ventures has also emerged as a potential segment of this field.

Future Prospects

India has a better scope in the future as there is a shift in trends towards tropical and traditional flowers and we can gainfully exploit this with our highly diverse indigenous flora. Specific attention is to be given to the development of traditional flowers by assisting traditional farmers in terms of bank loans, proper marketing facilities, and dissemination of information through training and media support on improved varieties. A hi-tech floriculture industry can be successful only if enough support is provided for procurement of genuine planting material either domestic or global and marketing of production by the government. Present status and growing trade is still in the early stages. Floriculture and Seeds in India are together being viewed as a high growth Industry but again to keep an eye on market, Flora expo is playing important role in creating market intelligence (Naqvi, 2011). Commercial floriculture has gained importance over the years. Technical collaborations by India with foreign companies is required to increase total share in the floriculture world trade.

Professional qualification combined with an inclination towards gardening and such other activities produce efficient floriculturists and landscaping professionals. The skills and knowledge required are imparted under the professional courses of floriculture and landscaping. The following measures, if taken will immensely benefit the entrepreneurs engaged in floriculture:

- Streamlining of export/import regulation, and provision of in-house quarantine certification.
- Creation of infrastructure for export like fleets of refrigerated trucks, cold storage at airports and pre-cooling facilities etc.
- Development and enforcement of quality standards.

- Allocation of guaranteed cargo space.
- Incentives to flower exporters-freight subsidy & cash compensatory system.
- Extensive international marketing and promotional support.
- Simplifying export documentation and billing procedures.
- Extending insurance coverage to include delayed delivery.
- Establishment of separate export promotion council on floriculture produces.

With increasing flower demand, production centers have expanded from traditional centers (USA, Japan, the Netherlands and Columbia) to new ones (Latin America, Africa and Asia). In Asia, India, China, Vietnam and Sri Lanka are moving in direction of intensive floriculture. New consumption centers are also emerging in South East Asia, Middle-East and Eastern Europe. This the moment to capitalize on our strengths and convert weaknesses into opportunities. However in the consumer-driven market, quality of service and delivery system will play a significant role. Marketing is no more a meeting of buyers and sellers. It is complex, driven by quantity of products and reliability of delivery. A new distribution structure with the aid of IT is likely to be in place, which would be a virtual market wherein demanding, branding, cataloguing and quality assurance would be emphasized. To be honest, most farmers are innocent and require timely supply of quality planting material and cultivation tips. Tighter checking is required to catch fake projects. The government should supply continuous electric power and look into providing of more concession. They should also have a control on the Value added tax (VAT) on cut flowers (presently 9-12%), and work at solving marketing problems. The government for building up the basic infrastructure of floriculture industry should offer loans with subsidies.

The main problem in floriculture is over-production, with quality being overlooked, resulting in low price realization. A large number of farmers are financially weak and uneducated. They do not understand which flowers are in demand in the market and what quantity they should grow. In addition, the growers have very poor know-how of technical packaging of flowers. The government clearly seeks to promote and develop floriculture in the state with active participation of local hill growers and entrepreneurs through latest technological intervention and thereby creating more opportunities for augmenting employment and income. However, some experts feel that the state can make much headway if schemes and strategies, worked out in workshops and conferences held at regular intervals, are implemented with a greater sense of urgency.

References

Arora, J.S. (2001). *Introductory Ornamental Horticulture,* 3rd ed. Kalyani Publishers, New Delhi, p.205.

Bhattacharjee, S.K. (2006). *Ornamental Crop Production Technology,* Pointer Publishers, Jaipur, p.344.

Chadda, K.L. and Chowdhary, B. (2007). *Ornamental Horticulture in India,* I.C.A.R., Kalyani Publishers, New Delhi, p.217.

Desh Raj. (2005). *Floriculture in Hills.* Agro Tech Publishing Academy, Udaipur. p.200.

Indian Horticulture Data Base. (2010). NHB, Gurgaon, p.22.

Ladha, S. (2011). IHITC technology route to create strong market intelligence. *Floriculture Today,* 15 (11): 10-12.

Misra, K K, S Kumar Lal and Srivastava, R. (2008). *A Practical Manual of Ornamental Horticulture.* G.B.P.U.A.&T., Pantnagar, p.84.

Naqvi, S.J. (2011). Exhibitions like Flora Expo and Horti Expo are the need of hour. *Floriculture Today,* 15 (9): 16-17.

Singh, G. (2011). Floriculture development in NE and Himalayan states. *Floriculture Today,* 15 (8): 18-22.

Tyagi, S. (2011). Grappling with problems through plant material suppliers. *Floriculture Today,* 15 (3): 12-15.

Our Contributors

Walter Fernandes is currently the Director, North- Eastern Social Research Centre, Guwahati, Assam and former Director of Indian Social Institute (ISI), New Delhi. Dr. Fernandes is an internationally reputed authority on the subject of Displacement, Resettlement and Rehabilitation issues in India and worked in the area for more than two decades. He has published several books and articles in edited volumes on the subject. He has been contributing papers to reputed national and international journals.

Uttam Lal works as Assistant Professor in the Department of Geography & Natural Resources Management, Sikkim University. His areas of academic interests include Himalayan environment, economy and culture. He has published research articles and also presented research papers in national and international seminars.

Swati Akshay Sachdeva works as Assistant Professor in the Department of Social Systems and Anthropology, Sikkim University. Before joining Sikkim University, she taught at Sikkim Govt. Law College for eight years and later in Sikkim Govt. College, Tadong in the Dept of Sociology. Her areas of research and teaching include Women Studies and Social Anthropology.

Binu Sundas, is an Assistant Professor in the Department of Social Systems and Anthropology, Sikkim University and is currently the Department Co-ordinator. He is the author of the book *"HiV and AIDS in Darjeeling: An Emerging Socio-Political Challenge"* (Concept Publishing Company, New Delhi) and *A Heady Mixture: Culture, Economics, Health, and Alcohol in Darjeeling* (Lambert Academic Publishing, Germany). His areas of specialisation are sociology of health and research methodology.

Deepti Singh works as Assistant Professor in the Department of Floriculture and Horticulture Management, Sikkim University. Before joining Sikkim University, she worked as Senior Research Fellow in Govind Ballabh Pant University of Agriculture & Technology, Pantnagar. The areas of her research have been heterosis breeding and combining ability in marigold.

Sohel Firdos is working as an Assistant Professor in the Department of Geography and Natural Resources Management, Sikkim University. On completion of his PhD. from the Centre for the Study of Regional Development, Jawaharlal Nehru University, New Delhi, he joined Centre for Studies in Social Sciences, Calcutta (CSSSC) as a Fellow in Development Studies in 2002. His current work relates to issues such as urban governance and socio-spatial exclusion in cities. He has travelled extensively in India and abroad to present his research. His research articles have found place in a number of academically acclaimed journals and edited volumes.

Otojit Kshetrimayum is a sociologist by training. Presently, he is a faculty member in V.V. Giri National Labour Institute, Noida. He is the Associate Coordinator of the Centre for North East Research and Training, and Associate Editor of the journal *Labour and Development.* His areas of interest include Cultural Studies, Sociology of North East India, Gender, Labour and Employment, Social Security, and Korean Studies.

Marchang Reimeingam holds PhD in Economics from the Centre for the Study of Regional Development, Jawaharlal Nehru University, New Delhi. Presently, he is a member of the faculty in the Department of Economic Studies and Planning, Sikkim University (SU), Gangtok.

Abdul Wahid Jasra is Rangeland Specialist and Team Leader Rangeland Resources Management (RRM) in the Environmental Change and Ecosystem Services (ECES) Programme of the International Centre for Integrated Mountain Development (ICIMOD). He has over 20 years of experience in the field of rangeland resources.

Srijana Joshi is a Consultant in ICIMOD. She is a plant ecologist by profession. She holds a PhD on evolutionary ecology of invasive plants from Tuebingen, Germany. Currently she is working in RRM and her research interest focuses on climate change and the economic valuation of ecosystem services.

Muhammad Ismail is Research Officer and team member of RRM in the ECES Programme in ICIMOD. He has over 12 years of experience in the field of rangelands.

Abdul Hannan is an Associate Fellow in Sikkim University. Earlier to this he worked as Programme Officer (Research and Monitoring) at Centre for Education and Communication, DFID, New Delhi. His areas of interests include Plantation Economy and Regional Development; Co-operative Farming and Microfinance; Fair Trade, Livelihood Studies and Sustainable Development.

Santosh Kumar Rai currently works as Associate Fellow in Sikkim University. He is a botanist by training and has published several articles on the subject. He specialises in ethnobotany of the Eastern Himalayas.

D. R. Chhetri is an Assistant Professor at the Post-Graduate Department of Botany, Darjeeling Government College. His research interest includes Himalayan Medicinal Plants, Ethnopharmacology, Plant biochemistry and Biotechnology. He has authored a number of research articles in reputed national and international journals and edited a couple of books.

Index

A

Acharya, 227
Ailments and treatment, 277
Amibhavi village, 141
ANC, 175
Appasamy, 132, 138
Arts and Crafts institute, 265
Assam, 24
Atharvaveda, 273
Ayurveda, 274

B

Basic needs, 134
Bhat, L.S., 141
Bhutan, 6
Bhutia women (of Sikkim), 211
    case studies, 229
    in decision-making, 227
Bhutia,
    introduction, 215
    origin and migration of

Bihu festival, 24
Biodiversity,
    and National Medicinal
        Plants Board, 105
    and CAMP workshop, 108
Boserup, 219
Boyer, Robert, 139
BPO, 173
Brahmaputra valley, 5
British East India Company, 170

C

Cameron, 143, 144
Cheitharol Kumbaba, 249
Chhetri, Dhani Raj, 91
Climate change, 29, 30
CSW, 173,
CTMI, 275

D

DACC, 181, 182
Dalits, 131
Darjeeling Himalaya,
    location, 91
    medicinal plant biodiversity
        in, 91-115
Darjeeling Himalayan region,
    traditional healing practices
        in, 273-283

Darjeeling,
  HIV/AIDS in, 169-190
    historical context, 171
  PLWHA in, 180
  response to HIV/AIDS in, 181
Data Highlights, 153
De Haan, 117
Decision-making, 224
  access to financial resources, 226
  children related, 225
  freedom from threat, 225
  mobility, 225
Deshpande, C.D., 141
DGHC, 171
Disasters, 22
  in northeast, 124
  search for solutions, 26
Dzongu valley (North Sikkim), 58
  demise of large cardamon in, 59
    causes of, 59

E

East India Company, 260
Eastern Himalayas,
  environmental degradation, climate change and disasters in, 22-28
Education,
  access to, 204
Engels, 212
Exclusion,
  and macro-economic development strategy, 139
  caste-based, 126
  from employment, 123
  from goods and services, 133
  from human rights, 138
  from land, 136
  from security, 137
  from social rights, 132
Fernandes, Walter, 22
Firdos, Sohel, 116
FRLHT, 108

G

Ganges system, 7
Ghurye, 121, 128
GJM, 171
GNLF, 171
Greenwood, Michael, 157

H

Handloom industry (in Manipur),
  case study of Imphal East District, 266
  under colonial patronge, 260
  under state patronage, 265
Hannan, Abdul, 191
Hasnain, Syed Iqbal, 52
Healing practices,
  in Bhutia, 275
  in Gorkha, 276

in Lepcha, 276
Himalaya,
climate and natural vegetation, 2
biogeograpphy of, 5
cultural setting, 9
eastern, 8
environmental and human dynamics, 1-21
economic arrangement, 13, 14
irrgated area, 16
land soil resource, 9
major ethnic composition, 12
water resource, 6
Hindu society, 121
Hindukush-Himalayas (HKH),
pastoral communities, 34
rangeland dependent mountain communities under climate change in, 29-46
HIV/AIDS, 169, 170, 171, 172
scenario in Darjeeling hills, 175
Hutton, 219

I

ICF, 270
ICIMOD, 30
IDU, 174, 176
Inclusion,
and exclusion, 144
India, 6, 93, 123, 125
understanding spatial and social exclusion in, 116-147
Indian Cooperative Societies Act, 1912, 205
International Labour Organisation, 116
IPCC, 47, 48, 60
IRCS, 175
Ismail, M., 29
Ives, J.D., 11

J

Jasra, A.W., 29
Joshi, S., 29

K

Kalichman, 169
Kallgren, 219
Kashmir valley, 13
Khawas, Vimal, 1, 47
Khelchandra, 239
Kinnaur,
distribution of agricultural land, 83
location, 70, 71
Klein, 30
Kshetrimayun, Otojit, 237
Kundu, 149

L

Lal, Uttam, 70
Lapche, 223
Lei marel, 247
Leinunglon, 247
Lhota Nagas, 219

M

Maharashtra, 137
Mandavlhare, S.M., 136
Manipur Handlom Industry, 264
Manipur,
  handloom weaving in, 237-272
  craft of weaving, 246
Marchang, Reineingam, 148
Marup, 265
Marx, 212
Medicinal plants, 99
  need to conserve, 100
  threatened, 103
Meiteis, 238, 239, 241
Migrants,
  concepts and definition of, 152
Migration,
  from NER to rest of the country, 148-168
  patterns and trends of, 155
  socioeconomic factors for, 149
Mountain biodiversity,
  and climate change, 103
Myrdal, Gunnar, 123

N

NACO, 175
Naga Chaddar, 248
National Campaign for Housing Rights, 134
National Medicinal Plants Board, 105
National Sample Survey, 123, 134
Nayak, 117
Nepal, 6
NHB, 284
Ningthonphi (royal aparel), 257
Niranjana, 213
Northeastern region (NER), 149, 150

O

Oba, G., 44
OBCs, 130

P

Paliteo, 291
Papola, T.S., 124
Pathak, Anjali, 55
Paul, 125
PLWHA, 170, 174, 177, 181
POB, 153, 154
POLR, 153, 154
Prasad, K., 23
Private institutions,
  policy inclusion of, 205
PRU, 125

R

Rai, Santosh Kumar, 273
Right of Children to Free and Compulsory Education Act, 209, 191, 195
Rigveda, 273

S

Sachdeva, Swati Akshay, 211
Seetharam, K., 50
Sen, Amartya, 117, 118, 123
Shah, 131
Shamanism,
  and people, 279
  and treatment, 280
Shanker Foundation, 178
Sikkim Himalaya,
  climate change and large cardamorn farming in, 47-49
  climate change in, 50
  glaciers in, 52
  large cardamom in, 53
    depletion and destruction of in, 55
Sikkim, 215, 227
  Bhutias of, 211-236
    women in work and decision-making, 218
  right to free and comulsory education in, 191-210
    dropont rates, 198
    physical infrastructure, 202
    teacher-pupil ratio, 199
    trained manpower in, 197
Singh, Deepti, 284
Social exclusion, 116-140
  and entitlement, 118
  due to poverty, 121
Social inclusion, 143
Somayaji, Sakarama, 1
South Asia, 1
Spatial exclusion, 140
  based on caste, 140
  in urban areas, 142
Srinivas, 129
State Advisory Council, 209
Sudershan, M.R., 61
Sundas, Binu, 169

T

TMAL, 275
Todaro, 166
Traditional dress,
  and social messages, 238
  hill-plain relationship, 245
  symbol and talisman, 239
TRU, 125

U

Umanglai Haraoba, 248
UNDP, 137
Unemployment, 119
UPR, 155
Uttar Pradesh, 130, 131, 141
Uttarakhand,
  floriculture in, 284-292
    future prospects, 281
    scope, 287

V

VCCTC, 175

W

WBSAP, 175

Weaving,
feudalisation of craft, 253
gender construction of, 258
under royal patronage, 248
West Bengal, 137
Western Himalaya,
environmental constraints and population pressure in mountain ecosystem of, 70-90
Women, 211, 212, 213
World War II, 264

Y

Yuthong Nyingthig, 275